Nanocomputers and Swarm Intelligence

Nanocomputers and Swarm Intelligence

Jean-Baptiste Waldner

First published in France in 2007 by Hermes Science/Lavoisier entitled "Nano-informatique et intelligence ambiante"
First published in Great Britain and the United States in 2008 by ISTE Ltd and John Wiley & Sons, Inc.

Apart from any fair dealing for the purposes of research or private study, or criticism or review, as permitted under the Copyright, Designs and Patents Act 1988, this publication may only be reproduced, stored or transmitted, in any form or by any means, with the prior permission in writing of the publishers, or in the case of reprographic reproduction in accordance with the terms and licenses issued by the CLA. Enquiries concerning reproduction outside these terms should be sent to the publishers at the undermentioned address:

ISTE Ltd
6 Fitzroy Square
London W1T 5DX
UK

www.iste.co.uk

© ISTE Ltd, 2008
© LAVOISIER, 2007

John Wiley & Sons, Inc.
111 River Street
Hoboken, NJ 07030
USA

www.wiley.com

The rights of Jean-Baptiste Waldner to be identified as the author of this work have been asserted by him in accordance with the Copyright, Designs and Patents Act 1988.

Library of Congress Cataloging-in-Publication Data

Waldner, Jean-Baptiste.
  [Nano-informatique et intelligence ambiante. English]
  Nanocomputers and swarm intelligence / Jean-Baptiste Waldner.
     p. cm.
  Includes bibliographical references and index.
  ISBN-13: 978-1-84821-009-7
  1. Molecular computers. 2. Nanotechnology. 3. Quantum computers. 4. Intelligent agents (Computer software) I. Title.
  QA76.887.W3513 2008
  006.3--dc22
                                                                2007045065

British Library Cataloguing-in-Publication Data
A CIP record for this book is available from the British Library
ISBN: 978-1-84821-009-7

Printed and bound in Great Britain by Antony Rowe Ltd, Chippenham, Wiltshire.

FSC
Mixed Sources
Product group from well-managed forests and other controlled sources
Cert no. SGS-COC-2953
www.fsc.org
© 1996 Forest Stewardship Council

Table of Contents

Foreword. ix
Didier TRUTT

Preface xiii

Acknowledgements xix

Introduction. xxi

Chapter 1. Revolution or Continuity? 1

1.1. Ubiquity and pervasion. 1
1.2. From the art of building small – perspectives of nanoproduction 4

Chapter 2. The Rise and Anticipated Decline of the Silicon Economy . . . 7

2.1. 40 years of global growth 7
2.2. From sand to the chip, the epic of semi-conductors 9
   2.2.1. Semi-conductors – some solid-state physics 10
   2.2.2. CMOS technology and high-density integrated circuits 14
   2.2.3. Half a century of industrial methods and processes. 16
2.3. The fatality of Moore's Law: "the wall". 22
   2.3.1. The disaggregation of the microelectronics industry value chain. . 26
   2.3.2. The ITRS (International Roadmap for Semi-conductors) –
   a race to controlled miniaturization 29
   2.3.3. Will a slowdown in the economy overturn established models? . . 33
2.4. Beyond silicon – from microelectronics to nanotechnologies 36

Chapter 3. Rebuilding the World Atom by Atom 41

3.1. Manipulation on an atomic scale – the scanning
tunneling microscope. 41

3.2. From the manipulation of atoms to nanomachines – the concept
of self-assembly... 45
3.3. From the feasibility of molecular assemblers to the creation
of self-replicating entities...................................... 49
3.4. Imitating nature – molecular biology and genetic engineering...... 55
   3.4.1. When nature builds its own nanomachines................. 56
   3.4.2. Genetic engineering – the nanotechnology approach
   by life sciences... 60
3.5. From coal to nanotubes – the nanomaterials of the Diamond Age ... 62
3.6. Molecular electronics and nanoelectronics – first components
and first applications... 70
   3.6.1. Carbon Nanotube Field Effect Transistors (CNFET)......... 71
   3.6.2. Hybrid mono-molecular electronic circuits................ 73
   3.6.3. Organic molecular electronics............................ 75
   3.6.4. Spin valves and spintronic semi-conductor components 82
   3.6.5. Quantum dots and the phenomenon of quantum confinement... 85

Chapter 4. The Computers of Tomorrow 89

4.1. From evolution to revolution 89
4.2. Silicon processors – the adventure continues.................. 93
   4.2.1. Progress in photolithography and new materials 95
   4.2.2. The structure of microprocessors........................ 101
   4.2.3. Digital signal processing and DSP processors............ 107
4.3. Conventional generation platforms............................. 109
   4.3.1. Traditional platforms................................... 110
   4.3.2. Emerging platforms 113
   4.3.3. Distributed computing, an alternative to supercomputers .. 115
4.4. Advanced platforms – the exploration of new industries 119
   4.4.1. Quantum information systems............................. 121
   4.4.2. DNA computing... 130

Chapter 5. Elements of Technology for Information Systems
of the New Century 135

5.1. Beyond processors.. 135
5.2. Memories and information storage systems..................... 138
   5.2.1. Memories consisting of semi-conductors – perspectives.... 140
   5.2.2. Limits of magnetic data storage 146
   5.2.3. Holographic memory...................................... 150
   5.2.4. The technology of AFM memories 154
   5.2.5. Molecular memory.. 156
5.3. Batteries and other forms of power supply 157
   5.3.1. Lithium-ion and lithium-polymer batteries............... 159
   5.3.2. Zinc-air batteries...................................... 160
   5.3.3. Micro-batteries... 161

  5.3.4. Micro-batteries using nuclear energy 162
  5.3.5. Recharging batteries with the help of kinetic energy 163
 5.4. New peripheral devices and interfaces between humans and machines 163
  5.4.1. Automatic speech recognition 165
  5.4.2. Gesture recognition. 170
  5.4.3. Processing and recognizing writing 171
  5.4.4. Eye tracking 172
  5.4.5. Brain machine interface 173
  5.4.6. Electronic paper. 177
  5.4.7. New visualization systems 181
 5.5. Telecommunications – a different kind of revolution 184
 5.6. The triumph of microsystems 187
 5.7. Is this the end of the silicon era? 190

**Chapter 6. Business Mutation and Digital Opportunities
in the 21st Century** 197

 6.1. Towards a new concept of information technology 197
 6.2. Ubiquitous information technology and the concept
 of "diluted" information systems. 199
 6.3. Highly diffused information systems – RFID. 204
  6.3.1. The "Internet of things" and the supply chain of the future –
  Auto-ID 209
  6.3.2. Economic opportunities vs. privacy protection 214
 6.4. New challenges for web applications in a global network of objects . 218
  6.4.1. Complexity and efficiency of very large infrastructures 219
  6.4.2. From centralized intelligence to swarm intelligence –
  reinventing the programming code 224
 6.5. The IT jobs mutation 231
  6.5.1. New concepts in agile software development 235
  6.5.2. Ambient intelligence and the delocalization of jobs
  in the IT sector 241
  6.5.3. New opportunities for the profession 245

Conclusion. 253

Bibliography 259

Index 263

Foreword

"We need men who can dream of things that never were." It is this sentence spoken by John Fitzgerald Kennedy in 1963 which has undoubtedly influenced my industrial career. Whether it be everyday electronics for the general public, or the components industry or even the world of video and imagery, the technological innovations which have come into use these last few years have greatly modified the competitive market and ways of living. Some scientists talk of technological tsunamis because of the resulting economical and social stakes, which are so high. For the industrialist the main challenge is size, and the idea is to continually innovate, creating smaller and smaller objects with improved performances. Combining research with industry has no longer become a necessity but rather a real race against the clock in order to remain a leading force in this domain. Development strategies, conquering market shares, increases in productivity and ground-breaking ideas are now more than ever linked to the coordination of tasks which join research and industry.

The most staggering technological innovations of these last few years (those which have led to real changes in the job market, and those which have changed our everyday lives) were not based on the technology available at the time but were based on anticipation, i.e. looking ahead into the future. There are some striking examples that come to mind: from the era of the video recorder, today we have all adopted the DVD player in record time discovering video on demand and downloads, which have opened the way towards the dematerialization of physical supports. From the era of silverhalide photography, today we are all familiar with digital photography, using the computer as the central storage base and the central base for exchanging our documents. Despite what anyone says, we have entered into a virtual world where it is no longer a question of physical support but of speed and bandwidth. All of this would not have been made possible if certain scientists had not dared to think differently.

It could be believed that once an innovation is launched and in the time that is taken for its economic democratization as well as its acceptance by consumers, researchers and industrialists would have the time for a little break. However, this is not the case. As soon as the technological innovation appears on the market, we are already dreaming of a new, more advanced innovation which will be one stage more advanced in terms of speed, ease of use, comfort and cost. We want to make all of this possible while also reducing the size of the object. We are looking for a groundbreaking opportunity which will open the doors of a new world to us. We are dreaming of a nanoworld.

The nanoworld is the world of the infinitesimal with exponential capacities and unexploited possibilities. Let us imagine a world where technology is no longer what it currently is. Processors have disappeared, replaced by tiny living molecules capable of managing large amounts of data; these molecules are also self-managing and have the ability to recreate themselves. A world where the alliance of biology, chemistry and physics would lead to advances in technology, which up until now have been unimaginable, like dreams or science fiction. This world is not far away. The beginnings are already taking place; the largest nations and the largest research laboratories are already regularly working on nanotechnology.

Imagine what the impact on industry would be. Will nanotechnology be the object of all our desires? Will it modify our economic models and our ways of life? Will it change the competitive market forever? Will industrialists be able to manage the change in jobs which will be brought about by this new technology? We are unable to predict whether or not by 2020 the innovations arising from the arrival of nanotechnology are going to restructure all sectors of industry. Will we be dressed in nanoclothes with unfamiliar properties? Will we travel around in nanocars with extremely resistant materials and pneumatics with an ecological engine and with renewable energy? Will production factories become molecular factories where we will see nanorobots? One thing is certain: nanotechnology is going to bring considerable improvements to the fields of computing, biotechnology and electronics. Thanks to more resistant and lighter materials, smaller components, new data storage systems and ultra-rapid processing systems, human performances are going to increase. We start off with simple isolated molecules and basic building blocks and end up with staggering technological advances which open up a new horizon, a new world – the nanoworld.

From this point on, the issue of the management of information combined with modern technology is going to expand. How are we all going to manage this way of life in our professional lives as well as in our everyday lives? How are we going to prepare for the management of new, more open information systems? It is at this level that the senior executive role becomes more and more important. The managers of today will become the managers of tomorrow by managing new ways of thinking, new working tools and new applications. In all sectors, computing, industry or even services, managers will have to accept the challenges that size represents. They will be the key players in the transition towards the nanoworld, they will be the founders of the revolution.

Didier TRUTT
Senior Executive Vice-President of the Thomson Group

Preface

"In the year 2000 we will travel in hypersonic airplanes, people will live on the moon and our vehicles will be fueled by atomic energy." As young children in the 1960s, this is how we imagined this symbolic milestone.

We believed that the 21st century would open a bright future brought about by beneficial technology, technology which had already been introduced to us at the time.

Yet, the year 2000 has now passed. The 21st century is here but this vision, which seemed to be shared by many people, almost belongs to kitsch iconography. Things have not changed in numerous areas: we still drive vehicles equipped with internal combustion engines, whose technology is more than 110 years old, airplanes travel once again at more economical speeds, there are energy crises, and the fear which nuclear energy raises has restricted any initiative for new applications – although high-speed trains draw their energy from nuclear fission.

The end of the 30 year boom period after World War II marked the end of major innovative and exploratory programs: the conquest of space was limited to the nearby planets, while Aerotrain, Concorde and other vertical take-off and landing airplanes were sacrificed for profitability. It became difficult to distinguish between scientific and technical vocations. Would technological perspectives become blocked? Would our research laboratories and departments become short of productive and profitable ideas in order to exclusively invest in quantative hyper-productivity that some people are already interpreting as destined to compensate for a lack in innovations?

However, ignoring the technological progress of the last 40 years would be iniquitous: the evolution of computing and of digital technology is conclusive evidence of the progress which has been made. From a few dozen reliable machines and with relative progress, we have entered into a market which is made up of billions of units. Components of machines have been reduced in size by a factor of 10,000 and their cost has decreased by the same proportions (a gigabit of memory is 5,000 times less expensive today than it was in 1970). Computers are everywhere nowadays, to the point where we no longer notice them when they are present.

Progress (the most envious industrialists will talk of simple industrial development) has continually been made by exploiting and perfecting the methods and processes of the most capitalist sector in history: the silicon industry. This is an industry whose evolution seems so predictable, so inevitable, that it has been characterized by a 40 year old exponential law, Moore's Law, which states that the number of transistors that make up microprocessors doubles every two years because these transistors become twice as small in the equivalent timescale.

However, this growth cannot go on forever and at this rate the limit of the performances of silicon components will be reached in 10-12 years at most. At that time, the current manufacturing processes will have arrived at the extremes of the world of traditional physics, the world of ordinary objects possessing mass, size, shape and contact properties. We will have arrived at a domain where the laws of quantum physics reign, laws which are so out of the ordinary and so different that they will call into question our industrial and technical know-how.

What will then be the perspectives for the development of our computers? What will the world of computing become in 15 years' time?

The anecdote about the year 2000 reminds us how dangerous it is to make this type of prediction. Nevertheless, this issue is far from being a simple puerile game. In numerous sectors of the economy and for many companies, 10, 12 or 15 years constitutes a significant time reference for the world of computing. Today, for example, the development process of the pharmaceutical industry considers information systems as the key to success in the development of new medicines, which is expected to have a project plan of between 10 and 15 years. This means that economic success will rest on the ability to understand and domesticate trends and innovations regarding technology, systems and organization. The aeronautical industry also considers project plans of 10 to 15 years and considers computing to play a vital role in this sector of the economy. We could also add the car industry and many other services such as banking and insurance for which information systems are no longer back-office production activities but rather real working partners. This timescale of 10 to 12 years also translates into more jobs available on the job market. We have only too often taken part in mass production campaigns in directing young people towards new industries, which five to seven years later proved to be too late in relation to the needs of the market.

As far as the next 10 years are concerned, by continuously pushing the limits of technology and the miniaturization of components further and further, it is the concept of the information system which changes: from processing, which is exclusively centered on the user, computing is becoming swarm intelligence. Since this technology currently makes it possible to produce tiny computers, it gives almost all objects from everyday life the capability of starting a spontaneous exchange of information without any interaction with the user.

In approximately 10 years computing will have become fundamentally different: it will not change gradually as has been the case for the last 15 years, but it will change dramatically for reasons which affect its deepest roots. Semi-conductors have been favored in the progress of computers over the last half-century. The simplest principle of thought is as follows: since it has been admitted that in 10-12 years, when hardware technology has evolved to its physical limits, then either our computers will have definitively reached their asymptotic power and future progress will be linked to innovative applications (this is the conservative or pessimistic view), or a replacement technology will emerge which will enable a joint continuation of hardware performance and progress in terms of application.

If the conservative point of view is acceptable on an intellectual and philosophical level, it will be difficult to accept for one of the most powerful sectors of the economy: to consider a ceiling in terms of the performance of components would mean entering into a market of hyper-competition where the only solution would be fatal price erosion. This is a situation similar to that of cathode-ray television screens and the textile industry, but with more serious consequences for the global economy given the volume of activity.

The hope therefore rests within replacement technology and the nanotechnology industry which will make the molecules itself. These molecules will be the base elements of future electronic circuits. For both economical and technical reasons, their development is considered as inevitable.

The first magnetic components (i.e. on the scale of the molecule where the rules of quantum physics reign) have already been created in laboratories. Conscious of the vital character of this technology, the large operators of the silicon industry are also amongst the leading pioneers of this domain. They are exploring carbon nanotubes in particular.

However, what can be created on the testing beds of laboratories on a unitary scale is still far from being applied to complex circuits and even less so as regards any possibility of industrial production. The main reason is the difficulty in the integration of several hundred million of these devices within the same chip like current microprocessors (remember that by 2015 there will be 15 billion transistors integrated onto one single chip).

Nanotechnology, in the historical sense of the term, does not simply consist of creating objects with molecular dimensions. The basic idea of the principle of self-organization into complex systems comes from life sciences. This means that the basic building blocks, according to the destination of the device, are capable of assembling themselves from just one simple external macroscopic instruction into a more complex device. This is the main challenge and the source of polemic debates[1] which scientists and technologists have to deal with today.

Given that nature can construct such machines and that molecular biology and genetic engineering have already investigated these principles, another path has opened for the computers of the future: that of organic molecular electronics which exploits a living material, such as a protein molecule, and reuses it in an artificial environment in order to ensure a processing or memory function.

The make-up of computers of the next decade is experiencing a transition period where everything is undergoing a process of change. New processors stemming from the conventional generation of semi-conductors are progressively going to be composed of structures coming from organic technology or from the first molecular components. Memory will follow the same path of miniaturization on a molecular scale. Mass storage devices will store information in three dimensions (in proteins or in crystals) which up until now has been stored in two dimensions, on the surface of an optical or magnetic disk. However, these magnetic disks, just like silicon components, will continue to be improved by exploiting quantum phenomena. Mobility has introduced another requirement: that of energetic autonomy. Chemical batteries will be produced flat, just like a sheet of paper, lighter and more powerful. The new generation of swarm and communicating devices will explore new approaches: supplying information using ATP molecules[2] (similar to living things), micro-internal combustion engines, or even nuclear micro-batteries which are far removed from conventional batteries. These swarm micro-objects, which are present in the diversity of the real world, are capable of communicating amongst themselves and with their users in the most natural way possible. This means that the interfaces

---

1 This debate amongst scientists hides another emerging debate concerning the general public. Just like GMOs, nanotechnology can be part of another controversial subject between researchers and the general public. The idea of designing tiny entities capable of reproducing themselves and which can escape from the control of their designers has been installing fear amongst people for some time now. The most futuristic practical applications illustrated in this book, which are part of current research trends, are still far from posing this type of risk. However, efforts to popularize these future applications must be undertaken if we do not want society to condemn emerging science through sheer ignorance of the real stakes and risks.
2 A reversible molecular turbine.

between the user and the machine will no longer be limited to keyboards, mice and screens but will use our five senses. Certain interfaces will not use the five senses since the neural system and telepathic control are already a reality. It is not the eye that sees, but the brain.

The vast network of tiny and heterogenous objects which make up the new generation of distributed systems enforces a new way of thinking in relation to software. Jobs in the world of computing will have to adapt to this. The development and maintenance of applications are no longer restricted to a finished set of programs with long lifespans, but have to take into account a vast perimeter of microsystems interacting with one another as well as unbelievable diversity and all in a context where instability and evolution rule, just like in the real world.

This book is devoted to the ambitious question of imagining what computing will be like in 15 years. How will information systems evolve and how will jobs in the computing industry change throughout this transition period?

Acknowledgements

IBM Thomas J. Watson Research Center, Yorktown Heights, NY, USA.

IBM Zurich Research Laboratory, Zurich, Switzerland.

Atomic Energy Commission, Quantronics Group, Saclay, France.

The author would like to thank the Reading and Validating Committee for their contribution that allowed all information to be presented in a correct and objective way:

Daniel Estève, Research Director at the CNRS. Mr. Estève received the silver medal of the CNRS and the Blondel medal for his work on semi-conductors and microelectronics. He worked as the Director and Deputy Director of LAAS/CNRS between 1977 and 1996. Between 1982 and 1986 he occupied the position of the director at the French Ministry for Research. Mr. Estève obtained his PhD in 1966.

Claude Saunier, Senator and Vice-President of the French parliamentary office on scientific and technological choices.

Didier Trutt, General Director of the Thomson Group and Chief Operating Officer (COO). Mr. Trutt graduated from the Ecole Nationale Supérieure des Mines in Saint Etienne, France.

Patrick Toutain, Senior Technical Analyst at the Pinault Printemps Redoute Corporation, graduated from the Ecole nationale supérieure de physique in Strasbourg, France.

Louis Roversi, Deputy Director of EDF's IT department (French electricity supplier), PMO Manager for EDF's SIGRED Project. Mr. Roversi graduated from Ecole Polytechnique, ESE, IEP, Paris, France.

Etienne Bertin, CIO of FNAC Group, graduated from the ESIA, DESS in finance at Paris IX, Dauphine.

Michel Hubin, former researcher at the CNRS and former member of a scientific team working on sensor and instruments at the INSA, Rouen, France, PhD in science and physics.

Thierry Marchand Director of ERP and solutions BULL, Managing Director of HRBC Group BULL, graduated from the EIGIP.

Introduction

We are on the brink of a huge milestone in the history of electronics and have just entered into a new digital age, that of distributed computing and networks of intelligent objects. This book analyzes the evolution of computing over the next 15 years, and is divided into six parts.

Chapter 1 looks at the reasons why we expect the next 10 to 15 years will be a split from conventional computing rather than a continuation of the evolutionary process. On the one hand, the chapter deals with the emergence of ubiquitous computing[1] and pervasive systems, and on the other hand it deals with the problem of the miniaturization of silicon components and its limits.

Chapter 2 offers a clear chronology of the industrial technology of silicon and explains the reason for the inevitable entry of nanodevices into the market: from Bardeen, Brattain and Shockley's transistor in 1948 to modern chips integrating tens of billions of transistors and the curse of Moore's Law.

For the reader who is not familiar with how a transistor works or who wants a brief reminder, there is an entire section which presents the main part of the theory of semi-conductors: elements of solid-state physics, the PN junction, the bipolar transistor and the CMOS transistor. The specialist does not have to read this section. This is also a section devoted to the current manufacturing processes of semi-conductor components. It is reserved for readers who wish to further their knowledge and is optional to read.

---

[1] Ubiquitous means micro-organisms that are found everywhere. Adapted to computing, the term refers to an intelligent environment where extremely miniaturized computers and networks are integrated into a real environment. The user is surrounded by intelligent and distributed interfaces, relying on integrated technologies in familiar objects through which they have access to a set of services.

The main part of the second chapter resides in the factual demonstration of the fatality of Moore's Law and the transition from microelectronics to nanotechnology (see sections 2.1, 2.3 and 2.4 for more information).

Chapter 3 is devoted to the introduction of nanotechnology and its application as the base components of computers.

First of all, it introduces the atomic microscope and the atomic force microscope, and the first experiments of the positional control of an atom with the help of such a device. The impossibility of constructing such a basic macroscopic machine by directly manipulating atoms is also explained. This would involve assembling billions of billions of atoms amongst one another and would take billions of years to complete.

The solution could reside in the self-assembly of nanomachines capable of self-replication. This is the approach of Eric Drexler's molecular nanotechnology (MNT) which is the origin of the neologism nanotechnology. A section is devoted to the polemic concerning the reliability of molecular assemblers in the creation of self-replicating entities (section 3.3). This section stresses the precautions which still surround the most spectacular and also most unexplored industries. Nanomachines are part of a real interest, especially as far as institutional circles (universities, research laboratories) are concerned. More and more leaders from the business world have been interested in nanomachines since they were first considered as a strategic sector in the USA. Section 3.3 is supplementary and if the reader opts not to read this section it will not affect their understanding and preconceptions relating to MNT.

The construction of self-replicating and artificial nanomachines is a topic which is still very much discussed. However, an alternative already exists: the approach used by living organisms and which has already been largely explored by molecular biology and genetic engineering. In fact, the powerful device that is nature enables the cell mechanism to assemble, with the perfection[2] of atomic precision, all varieties of proteins; an organized system of mass production on a molecular scale. Section 3.4 explores the construction of self-replicating nanomachines which are used effectively by nature. This section also develops another approach to nanotechnology which is derived from genetic engineering.

---

2 It would be appropriate to talk about almost-perfection. It is the absence of perfection in the reproduction of biological molecular structures which enables nature to be innovative and to become everlasting. Without these random imperfections, most of which reveal themselves as being unstable and non-viable, no sort of evolution would have been possible, and man would probably have never appeared.

Section 3.5 introduces carbon nanotubes and their electronic properties. These extraordinary materials, which would make a car weigh about 50 lbs, are the first industrial products stemming from nanotechnology.

At the end of this chapter, section 3.6 is devoted to the introduction of some remarkable basic devices which function on a nanometric scale and which are likely to replace the semi-conductor components of current computers. The theoretical and experimental reliability of computing which is based on molecular components are also presented. The CFNET (Carbon Nanotube Field Effect Transistor), hybrid mono-molecular electronic circuits, an organic molecular electronic device using a living protein to store a large amount of information and finally spintronic semi-conductors will all be dealt within this section.

Finally, quantum boxes and the phenomenon of quantum confinement will be discussed. This quantum confinement approach is currently more a matter for fundamental physics rather than the development of industrial computers.

Chapters 4 and 5 are meant to be a summary of the major technologies which the computers and systems of the next decade will inherit. These two chapters can be read in full or according to the interest of the reader. As the sections are independent, they can be read in any order.

Chapter 4 is devoted to processors and their evolution. This chapter introduces two analytical views. First of all, we address the standard outlook of between one and five years (microprocessor structure: CISC, RISC, VLIW and Epic, the progress of photolithography, distributed computing as an alternative to supercomputers, etc.), i.e. the perspective which conditions traditional industrial investments.

The chapter then introduces a more ambitious perspective which sees a new generation of computers with radically different structures. This vision with more hypothetical outlines introduces systems which may (or may not) be created in the longer term, such as the quantum computer and DNA processing.

Chapter 5 widens the technological roadmap from the computer to the entire set of peripheral components which make up information systems. It is structured according to the hierarchy of the base components of a system. Other than processors, we are also interested in memory, mass storage devices, dominating energy supply devices with the notion of mobility or distributed computing, and in the man/machine interface. There is a section which is specifically devoted to microsystems whose contribution to ubiquitous computing is vital. The technologies which have been mentioned above can be applied to the industrial present (notably with new semi-conductor memories, Giant Magnetoresistance (GMR) hard disks, and Tunnel Magnetoresistance (TMR) hard disks, voice recognition and new visual display devices) as well as future possibilities (holographic memories, memories created from atomic force microscope (AFM) technology, molecular memories and telepathic interfaces, etc.).

Chapter 6 deals with the changes that these new resources will introduce in the business world and in our everyday life. It takes into consideration the issues at stake in the business world: a break or continuity of investments, the impact on activities using coding, and the implementation of information systems in companies, economic opportunities and changes in business and jobs, etc.

Section 6.1 introduces the historical evolution of computing from the first specialized, independent systems through to ubiquitous systems.

Section 6.2 shows how we move from ubiquitous computing to an ultimate model of dilute computing (i.e. becoming invisible): tiny resources communicate with one another and together they resolve the same type of problem as a large central mainframe. It also shows how the relationship between man and machine has evolved over three characteristic periods.

Section 6.3 introduces one of the first applications considered as pervasive, a precursor to dilute systems: Radio Frequency Identification Systems (RFID) and the Internet. We will also show how this global network could structure the supply chain of the future which would concern the Auto-ID initiative. The section concludes with the sensitive issue of the potential attack on a person's private life which this type of application could lead to.

Section 6.4 introduces the new challenges of a global network, i.e. how it is possible to make such complex and heterogenous networks function on a very large scale. Interconnected networks, which would have billions of nodes, and which would be associated with so many diverse applications, infrastructures and communication protocols, pose three fundamental problems. First of all, there is the problem of security, secondly the problem of the quality of service and finally, the size or the number of interconnected objects. However, such a structure would also consist of moving from a central system to what is known as swarm intelligence. In order to effectively use these swarm networks as new applications, the rules of software coding would have to be reinvented.

Section 6.5 mentions the unavoidable change in businesses and jobs in the computing industry, which the arrival of these new structures will lead to over the next decade. These new software structures will completely change the way in which we design, build and maintain complex applications. The following issues will also be mentioned in this section: the modern concepts of agile development, new opportunities regarding professions in the computing industry and the sensitive issue of off-shoring since distributed structures make it possible to operate these activities in regions where costs such as those of employment and manufacturing are more attractive.

After an analysis of the evolution of the business world in the computing industry, as well as an analysis of the evolution of structures in the computing world, section 6.5 deals with what the essential reforms in jobs relative to information systems should be; what will the new professional industries introduced by this technological evolution be like, and finally what jobs will no longer be viable?

The book concludes by reformulating the reasons behind the joint emergence of the material revolution (post-silicon and post-nanometric devices) and the concept of swarm intelligence, i.e. a complete and simultaneous change which is not only present at the center of materials technology, but also in algorithms which will unite this collective intelligence and in the perception that the users will have of these new systems. This conclusion introduces the real theme of the book: processing the next generation of computers which will enable us to understand the impact of change that these machines will have on methods used by humans and the manner in which humans manage these machines. We benefit from the understanding acquired from available technology in order to be able to debate the necessary change of the CIO (chief information officer)'s role: a job which from now on is no longer development and production oriented, but oriented towards anticipation, vision, mobilization and integration.

Chapter 1

Revolution or Continuity?

1.1. Ubiquity and pervasion

At the beginning of the 20th century the use of electricity was confined to installations which were located close to a central power station. This use of electricity was limited to providing lighting for only several hours during the day. Nowadays, electricity is used in almost every domain and is available nearly everywhere throughout the world. For most consumers it is difficult to imagine the number of appliances powered by electricity.

For a long time sources of mechanical energy were only available to large steam engines. The first automobiles then started to use this energy. A sophisticated car currently has at least 30 electric motors which operate the seats, windows, roof, mirrors, etc.

The evolution of a particular type of technology is clearly linked to the evolution of its most important and basic performances. Its evolution is also linked to its accessibility and its ability to function in all areas of everyday life.

One challenge that scientists will face over the next 10 years when creating computer systems will be to develop systems that enable people to process more information in a quicker, more ergonomic manner, in other words, the ability to improve the interaction between man and machine in the most natural, anthropomorphic way possible. However, we want to have all this information available to us everywhere we go, and in any given situation. Every individual wants to be able to access their details, personal as well as professional, from anywhere in the world. Wherever they are or whatever their surroundings, these people want to be able to be contacted, receive their mail, obtain personalized information and instant responses to their different requests. Ideally, they would prefer to do this at any given moment and with wireless systems. This type of information is still evolving. With microsystems that are capable of working directly on site, the information we need is becoming more and more subject-related. Requirements in terms of processing power and storage space add another constraint: the increased switching capacity means an increase in the number of mobile computer systems whose size has been reduced to proportions which were up to now technically impossible to create. These distributed systems will be made available everywhere in the real world.

The structures of processing systems from the last century were based on batch processing, then on time sharing and transactional monitors operating on large systems. With the PC we saw the arrival of customer/server structures and with the Internet we saw the arrival of web solutions. The 21st century will open the doors to new solutions which will atomize data processing and standardize processing power.

These new structures which will be paramount in the very near future will change computers, information systems and their peripherals to the point where they will progressively be confused with the environment that they are supposed to control. Such systems will become part of our everyday lives in order to help our well-being and our security. People will barely notice them even though they will still use them. Who amongst us would have thought that by using a credit card with a chip during a commercial transaction, there would be a digital transaction between a tiny computer and the bank which issued the card? Who is concerned about knowing that when pushing the brake pedal of a car an instruction signal is sent to the ABS computer which decides the best way to use the brake pads in order to achieve the optimal stop? Omnipresent and often interconnected within large networks (such as geo-localization by GPS and the emission of an alarm signal in the case of an accident or the theft of a vehicle; range finding and mobile tele-surveillance), these microsystems which act directly on the physical world will become part of our normal everyday objects or even part of our own bodies.

From now on the simple keyboard and mouse will no longer be sufficient. The market demands more and more ergonomic applications. The availability of human/machine interfaces has been growing steadily over the last 10 years. The interfaces currently on offer have become more varied with voice recognition and the detection of movement used as a way of controlling certain types of applications. In this domain, the ultimate aim is to use telepathic commands. Why do our brains not directly control our computers? From here on this idea will no longer be simply an idea emanating from science fiction. Advanced research in the area of cerebral interfaces might already foreshadow future exchanges with computers. In order to understand the central place of the human/machine interface in computing and its phenomenal rate of evolution we just have to cast our minds back to the first word processed texts which displayed amber/green characters on a single line of a black screen. Less than 20 years ago we also witnessed the arrival of the predecessors to the computers that are used today with windows, icons, menus, and where everything is accessible at the click of a pointing device.

> The Alto system was the ancestor of the "mouse, menu, full graphic screen" PC which is, in fact, how all microcomputers are built today. The Alto microcomputer project produced by Xerox was developed by PARC engineers by reprogramming a mini 16 bit computer from that time period in order to achieve a graphic environment similar to that of the bitmap (meaning each point of the screen can be addressed by the operating system). Alto led to the creation of the graphic interface known as *SmallTalk*. The commands were transmitted by a revolutionary pointing device, known as the mouse. Alto was released in 1974 and became known as Xerox Star in 1981. It was first commercialized in Macintosh machines then on Windows PCs.

The classic reproduction of data on screen and on printers has also become too restrictive. Virtual reality, which for a long time was limited to the areas of teaching and leisure, is now being used more and more in industrial and professional applications. It is used in domains ranging from mechanical modeling to the reproduction of complex requests in very large databases. It is also used in synthetic vision devices such as NASA's AGATE program which is used in light aircraft and which provides a synthetic display of the visual field on a screen onboard the aircraft when unexpected weather conditions make the flight visually impossible (80% of general aviation accidents are due to a deterioration in visibility conditions). In our everyday lives there is also a desire to use a similar type of processing power and storage capacity. The processors of current video game consoles such as the XBox 360 and PlayStation 3 are as powerful as the supercomputers that existed about 10 years ago[1].

These systems are not only becoming more readily available and more ergonomic, but they are also becoming more powerful. Currently, the main aim of computers with increasingly powerful processing and storage capacity is to simulate and manipulate more complex models which increasingly resemble their real-life forms in all scientific and technical domains, from chemistry to aeronautics and from physical sciences to the automobile industry. It will be the norm for these future computers to integrate clock rates of several gigahertz, exceptionally high pass bands as well as high capacity, low-cost memories. However, these

---

1 With a processing power of 1 TFLOPS (10^{12} floating point operations per second) for the XBox 360 or 2 TFLOPS for PlayStation 3, games consoles are now formally classed in the category of supercomputers.

hypercomputers[2] will no longer be based on monolithic computers which have a steadily increasing unitary processing power. These computers will progressively be replaced by cluster systems, network systems or grid systems whose overall performance will have more of a mass effect than the single intrinsic processing power of the computers that make up each system[3].

Industrial and scientific researchers agree on the idea that in the medium term, the conventional technology used for semi-conductors will no longer be adapted to the design and production of these new, even more miniscule devices whose processing power will, from now on, be expressed in PetaFLOPS (1 PFLOPS = 10^{15} floating point operations per second). The FLOPS is the common unit which is used to measure the speed of processors. Over the next 10 years scientists anticipate the emergence of hybrid structures which will link the latest generations of semi-conductors with the precursors to the new technology used for the infinitesimal.

In the past, semi-conductor devices focused largely on transistors on a given silicon surface. The same can be said for storage density on magnetic disks. However, this process cannot continue indefinitely. It is essential to discover and understand the lifespan of these processes. It is also vital to understand the technological alternatives which are essential for a better understanding of computers for generations to come.

1.2. From the art of building small – perspectives of nanoproduction

Since semi-conductors have been the driving force behind the immense progress in computing throughout the last century, the question now is to find out exactly how well they can perform and how far they can be tested by scientists. Not very long ago, scientists considered that the integration of components of two microns was the threshold of absolute miniaturization for semi-conductor devices (the thickness of the circuit lines of the first Intel processors was 10 microns and it was thought to have been very difficult to cross the one micron barrier). Since the year 2000 this asymptotic vision has largely been overcome as scientists have created

---

2 IBM BlueGene/L with a theoretical processing power of 360 TFLOPS is no longer the world's most powerful supercomputer. It has been dethroned by MDGrape-3, developed by Riken in Japan. MDGrape-3 is a distributed system made up of a cluster of 201 units, each of which is made up of 24 custom MDGrape-3 chips. These 24 custom chips work with 64 servers, each of which is made up of 256 Dual-Core Intel Xeon processors and 37 servers made up of 74 Intel Xeon processors. This distributed supercomputer should be the first to pass the PetaFLOPS threshold.

3 Just as in physics, the processing power of systems is the product of the potential factor of the computers (the intrinsic unitary power of technology) with their intensity factor (the mass of intelligent devices working together in the framework of a global system).

transistors with dimensions reaching 0.18 microns. In 2004 state of the art processors were composed of circuits of 90 nanometers and from the first half of 2006 processors have been mass-produced with circuits 65 nanometers thick. Despite performance of such an impressive nature there is, however, an absolute limit at least for technology which stems from the conventional process of photolithography which will be discussed later in this book.

This book aims to explore, with the pragmatic vision of a financial and industrial analyst, alternative as well as conventional technologies (in other words, those coming from the evolution of current technologies such as extreme UV photolithography, hard X-ray lithography, electron beam engraving) and the more radical new approaches which nanotechnology offers when the traditional methods will have become outdated. We will see that all this is not a slow process of evolution taking place over several decades but rather a quick one which has been going on for approximately a decade.

> The more the processes of lithography evolve, the more the dimensions of transistors become smaller, enabling engineers to stock more of them on chips. The silicon industry has shown a successive change in the technology used and in the processes used in order to be able to industrially produce the smallest transistor. In this way, the most sophisticated technology in 1998 was able to create transistors of 250 nm. From 1999 the major manufacturers converted their most specialized production lines to 180 nm technology. This state of the art technology changed to 130 nm in 2003, then 90 nm in 2004, 65 nm in 2006, and 45 nm will be commercially viable in 2008. The 32 nm process is due to arrive in the 2009–2010 timeframe.

Today there is an abundance of literature, articles and reports written on the topic and the use of nanotechnologies by the popular press as well as more distinguished scientific and business publications. Yet too often these publications confuse the applications which will be available in our everyday lives in the very near future with those applications that belong to the more distant future or which are too hypothetical to be of any industrial or commercial interest.

Nanotechnology consists of producing smaller, quicker, more resistant objects as well as genuinely new objects or objects possessing properties which up until now have been non-existent. Nanotechnology also aims at producing machines which will introduce a new conception of production which will be radically different from anything we have experienced up to this point.

Do the uses of nanotechnology really constitute a revolution? This term has been overused in the industrial and commercial worlds in that the most insignificant of breakthroughs in any discipline is systematically qualified as a revolution. However, this time the term seems to be used in the correct context. The majority of scientists, industrialists and marketers recognize that a true revolution is going to take place. Perhaps this will be one of the largest scientific and technological revolutions of all time because it will affect absolutely all aspects of our daily lives in the following ways: from the medicines that we take to the nature of the materials which make up the objects present in our everyday lives, from our computers, telephones and other electronic products, not forgetting new sources of energy, the protection of the environment and the clothes that we wear, all contribute to the theme of this book. To sum up, nanotechnology constitutes a generic technology, a metatechnology whose advent will enable us to reform the majority of our industries in a radical new manner.

In the short or medium term, the fields in which nanotechnology is used in both science and industry will still remain very specialized, even if over a period of 15 years the most optimistic scientists were to see a convergence between all disciplines using nanotechnology towards a universal model that could embody *Molecular Nanotechnologies* (MNT). "Nanomania" (and in particular the substantial government funding granted to each discipline associated with nanotechnology) has led a lot of laboratories and companies to label everything influencing the domain of the nanometer as "nanotechnology" (from the production of nanostructured materials to life sciences such as genetics, including, of course, molecular electronics).

However, in this patchwork of domains which has no unifying link there is one unifying and multidisciplinary character that emerges: on the nanoscale each object is an assembly of the same elementary building bricks – the atom! On this scale at one millionth of a millimeter the physical, mechanical, thermal, electrical, magnetic and optical properties directly depend on the size of the structures and can differ fundamentally from the properties of the material on the macroscopic level. This has been the case experienced by scientists until now. This is due to a number of reasons which, of course, include the quantum behavior but also the increasing importance of interface phenomena. The importance between area and volume increases with the degree of miniaturization. In nanometric terms, this leads to a certain behavior of the surfaces of the circuit which then dominates the behavior of the whole material which the circuit is on. Nanostructures, in this way, can also possess many more mechanical or physical properties than there would be on the same material in its macroscopic form. As soon as scientists are finally able to master atomic scale precision, then these properties will be radically modified for optimal performance, thus creating completely new perspectives on an industrial level.

Chapter 2

The Rise and Anticipated Decline of the Silicon Economy

2.1. 40 years of global growth

The 20th century was the century of the electricity fairytale. Throughout its domestication, the boom in electrical engineering and that of the telecommunications industry marked the beginning of a continual period of economic growth.

With the internal combustion engine and the first phases of the automobile industry, electrical engineering was the main force behind the economic development of the Western world up until the end of World War II.

The second half of the 20th century saw a new shift in growth regarding electrical engineering, especially with the advent of the transistor which opened the way to the race to the infinitesimal.

The invention by Bardeen, Brattain and Shockley in 1948 and the introduction of binary numeral principles in calculating machines at the end of the 1930s paved the way for the silicon industry, which has been the main force behind global economic growth for more than 40 years now.

Since 1960, the size of electronic components has been reduced by a factor of 10,000. The price of these components has also collapsed in equivalent proportions; from $101,000 in 1970 you can now purchase a gigabyte of memory for approximately $27 today. No other sector of industry can pride itself on such productivity: reducing unitary production costs by a factor of one million in the space of 30 years. The story does not stop there.

· Digital microelectronics has become a pillar in virtually all sectors of industry: the automobile industry was considered as the "Industry of Industries" throughout the 30 year boom period after World War II in France, and more than 90% of new functionalities of modern cars come from microelectronic technologies. Today, electronics count for more than one-third of the price of a vehicle.

However, the most significant domain has, without a doubt, been the development of the personal microcomputer. With the arrival of large information systems we saw the birth of digital electronics. These sizable structures use large central systems whose usage had been very restricted. There were only a few specialists who worked in this field who really knew how to use these systems. We had to wait for the arrival of the PC for the general public to have access to the world of informatics and information technology. This mass generalization of laboratory-specialized technology for everyday use has created many industrial opportunities; this model, which is conditioned only by technical performance up until now, should, for the moment, benefit both cost and performance. Technology entered an era of popularization just like all other inventions: from electrical energy to the automobile and aviation. A new stage in this process was reached with the digital mobile phone. For the moment, no other economic domain can escape from this technology.

> Pervasive technology is technology that spreads on a global scale, technology which "infiltrates" or becomes omnipresent in all socio-economic applications or processes. Pervasive is a word with Latin roots and became a neologism towards the end of the 1990s. The word pervasive minimizes the somewhat negative connotation of its synonym "omnipresent" which can install fear of a hyper-controlled Orwellian universe. The word pervasive is reassuring and positive giving the notion of a beneficial environment.

A new chapter in the evolution of both science and technology has been written following a process which from now on will be referred to as being traditional. This evolution includes discovery, accessibility of such high-tech applications for the elite only, popularization of such high-tech applications to all social classes, the emergence of products coming from this technology and the growth of products that this technology will create. For example, with regard to the first main usage of electricity we automatically think of providing light for towns. The technology, which is available today, has become a more decentralized universal source of energy (the battery, miniaturized engines) and finally has become so widespread that it can be used for anything and be accessed by everyone.

Omnipresent electronics in our daily lives has become omnipotent in the economy with a growing share in global GNP. At the beginning of the 21st century, microelectronics represented €200 billion in the global GNP and in the silicon industry. This activity generates more than €1,000 billion in turnover in the electrical industry and some €5,000 billion in associated services (bringing the global GNP to some €28,000 billion).

The share of this industry has, in all likelihood, led to a further growth with the phenomenon of pervasive systems, which integrate more and more microelectronic technology into everyday objects. The silicon industry has probably got a good future in front of it and has no reason to be doubted over the next 12 to 15 years.

2.2. From sand to the chip, the epic of semi-conductors

Throughout the 20th century, the rate of evolution accelerated. The "bright future" promised by modern society after World War II stimulated technological progress, the productivity race and doing work that demands the least physical effort. The more complex machines are also the most powerful ones and are now available everywhere. Their designers now understand that the future of technology resides in the process of miniaturization.

One of the most basic of these machines, the switch, has benefited greatly from this process and is being produced in increasingly smaller sizes. In the 1940s, following electromagnetic relays, the triode enabled scientists to join a number of switches together in complex circuits. The electronic calculator had just been born.

> A very simplified description of the digital calculator would be to describe the device as a processor which carries out arithmetic or logic functions and has a memory that temporarily stores products and sub-products of these operations in the long term. The logic used for all the operations is based on the only two states that an electronic circuit can possess: 0 or 1. The switch is, in fact, a device which makes it possible to obtain an output signal of comparable nature to the input signal. The switch is to electronics what the lever is to mechanics. A lever makes it possible to have a greater force at the output than at the input, or even a greater (or weaker) displacement than that applied at the input. It is with the help of the basic switch that AND, OR and NOT gates can be created. These logic gates used in electronic circuits form the basis of all the operations carried out by our computers.

The 1960s were devoted to the advent of the transistor and semi-conductors. These new components progressively replaced cumbersome, fragile vacuum tubes which, at the time, were both time and energy consuming. Computers started to function with the help of what we can consider as chemically treated sand crystals. This was the arrival of silicon. Silicon is the eighth most abundant element in the universe, whereas on Earth it is the second most abundant element, coming after oxygen. Silicon has contributed to the development of humanity since the beginning of time: from flint to 20th century computers including the most diverse everyday ceramic objects.

2.2.1. Semi-conductors – some solid-state physics

All solids are made up of atoms that are kept in a rigid structure following an ordered geometric arrangement. The atom is made up of a positively charged central nucleus, surrounded by a negatively charged cloud of electrons. These charges balance each other out in such a way that the atom is electrically neutral. A crystal is a regular arrangement of atoms bonded together by a process known as covalence, i.e. the sharing of electrons. It is the bonds which give the solid its cohesiveness. The electrons, which take part in the cohesiveness of the solid, are bonded electrons. The excess electrons, if any exist, are said to be "free electrons". The detached electrons can travel freely around the interior of the solid. They allow for the easy flowing circulation of electrical current which characterizes materials with conducting properties.

On the other hand, insulators do not have any (or have very few) free electrons (we will talk about electrons in the conduction band) and in order to make an electron pass from the valence band (i.e. a bonded electron which assures the rigidity of the solid) towards the conduction band, a significant supply of energy must be supplied which will bring about the destruction of this insulator.

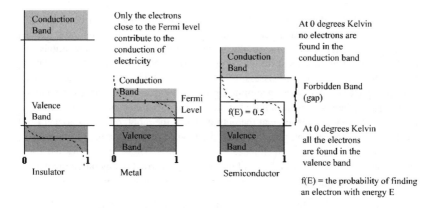

Figure 2.1. *The valence band, conduction band, forbidden band and the Fermi level*

In between insulators and conductors, there exists a type of solid which possesses a medium level of resistivity. These are known as semi-conductors such as germanium, selenium, copper oxide and of course silicon. In a semi-conductor, the conduction band is empty at the temperature of 0 degrees Kelvin: therefore, the semi-conductor is an insulator. However, the difference with an insulator here is that a moderate supply of energy can make the electrons migrate from the valence band to the conduction band without harming the body and without an increase in temperature. Whilst an electron is free, it leaves behind it an extra positive charge in the nucleus of the atom and also leaves a place available in the bond between two atoms.

This available place acts as a hole which a neighboring electron of a crystalline lattice would have the possibility of occupying, freeing yet another hole which would be occupied by a new electron and so on.

This is the phenomenon of the propagation of bonded electrons. Double conduction is also possible such as the double conduction of free electrons traveling in the conduction band and that of the propagation of holes/bonded electrons in the valence band.

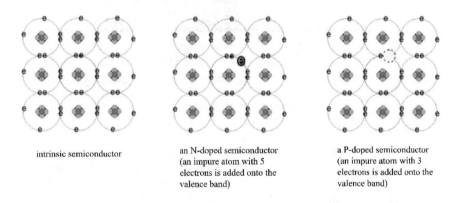

intrinsic semiconductor

an N-doped semiconductor
(an impure atom with 5 electrons is added onto the valence band)

a P-doped semiconductor
(an impure atom with 3 electrons is added onto the valence band)

Figure 2.2. *An intrinsic semi-conductor, an N-doped semi-conductor and a P-doped semi-conductor*

The number of free electrons in a crystal can be increased by adding an impurity made up of heterogenous atoms (one impure atom or one doping atom for one million atoms in the intrinsic semi-conductor).

Cohesion of the pure silicon crystalline lattice is obtained by the sharing of four valence electrons from the silicon atom. Imagine that some arsenic atoms, each one having five valence electrons, are introduced into the silicon crystal. One electron from each doping arsenic atom would find itself with no other electron to bond with.

These orphan electrons are more easily detached from their atoms than the electrons forming the bond between the atoms. Once free, these electrons favor electric conduction through the crystal. In reverse, if heterogenous atoms with only three valence electrons are added (for example indium atoms), then an electron is missing in one of the bonds which creates a hole in the crystal's structure, a hole which will also lead to the spread of an electric current. In the first case, we talk about an N-doped semi-conductor because there is an excess of valence electrons in the semi-conductor. In the second case, we talk about a P-doped semi-conductor because there are not enough valence electrons or holes in the semi-conductor.

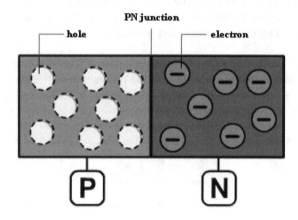

Figure 2.3. *A PN conductor*

The PN (or NP) junction is at the heart of current electronic components. The idea consists of creating an N-doped zone and a P-doped zone in the same semi-conductor crystal. Straddled on the junction that is created between both regions, a thin transition zone is also produced. Holes move from the P region to the N region and trap electrons. Electrons move from the N region to the P region and are trapped by holes.

This very narrow region is therefore practically devoid of free carriers (i.e. electrons or holes) but still contains fixed ions coming from the doping atoms (negative ions on the P side and positive ions on the N side). A weak electrical field is established at the junction in order to balance the conductor.

This field develops an electrostatic force which tends to send carriers to the area in which they are the majority: holes to the P zone and electrons to the N zone. When electrical tension is applied to the boundaries of the junction, two types of behavior can be observed: the crystal is a conductor if the P region is positive in relation to the N region, and is a non-conductor in the opposite case.

We say that the junction is "passing" from P to N and "non-passing" or "blocked" from N to P.

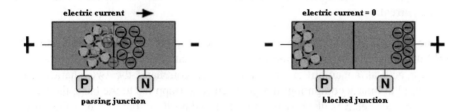

Figure 2.4. *A passing junction and a blocked junction*

By placing two NP and PN junctions side by side (or in reverse order) in the same crystal forming either two N zones separated by a thin P zone, or two P zones separated by a thin N zone, the device develops into a junction transistor.

With the emitter-base junction being polarized in the direction of the current flow, the barrier of potential E_{BE} is lowered and the crossing of dominant carriers is made easier. Electrons, which are the dominant carriers of the emitter, pass into the base.

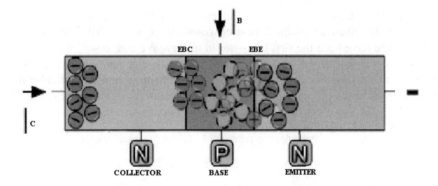

Figure 2.5. *A junction transistor (or bipolar transistor). The base is very thin and very slightly doped. There is very little combination of electrons and holes occurring here. The majority of electrons reach the collector, producing a collector current which is proportional to the current applied to the base. This is the transistor effect*

Electrons arriving in the base which is a P zone should combine with holes. This is not the case because the base is very thin and very slightly doped. Some combinations occur there but the majority of electrons have enough energy to reach the collector. It can be observed that the current in the collector is proportional to the current applied at the base. The transistor therefore acts as an amplifier of the electric current.

One example of how this device is used, and which is of particular interest to the world of digital electronics, is the use of the switching transistor. In effect, if the transistor acts as an amplifier, it is possible to command the two characteristic values used in binary computing; if a given current is applied to the base, the current in the transistor behaves like a switch in the closed position. In reverse, if the current that is applied to the base is zero, then the current in the collector will be almost at zero and the switch is therefore in the open position.

2.2.2. *CMOS technology and high-density integrated circuits*

Modern electronic circuits use a variant of the bipolar transistor described above. This type of transistor is in fact too cumbersome, too energy-consuming and more difficult to produce on a very large scale. In the race towards miniaturization, this transistor was rapidly superseded by a new device known as the Field Effect Transistor (FET) and in particular the Metal Oxide Semi-conductor Field Effect Transistor (MOSFET).

In 1930, J. Lilienfeld from the University of Leipzig applied for a patent in which he described an element bearing resemblance to the MOS transistor and which could have been the first transistor in history. We had to wait until the 1960s in order to see the arrival of such devices, whose development was made possible with the noted progress in the field of bipolar transistors and in particular in the resolution of problems with the oxide semi-conductor interface. Today, the MOS transistor is the key element of digital integrated circuits on a large scale due to both the simplicity of its production and its small size.

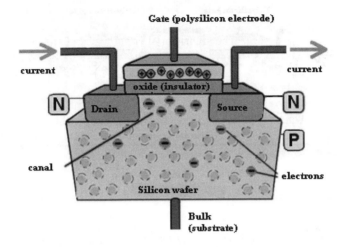

Figure 2.6. *The Metal Oxide Semi-conductor Field Effect Transistor (MOSFET)*

The MOS transistor is a FET made up of a semi-conductor substrate (B or Bulk) and covered with a layer of oxide on which there is the gate electrode (G or Gate). By applying a potential difference between the gate (G) and the substrate (B), an electric field in the semi-conductor is created causing the dominant carriers to be pushed back far from the oxide semi-conductor interface, leaving the non-dominant carriers coming from two complementary sectors in the substrate known as the source (S) and the drain (D). These sectors form a thin layer of mobile charges known as a canal. These charges are able to transit between the drain and the source, both of which are found at the far ends of the canal.

Complementary Metal Oxide Silicon (CMOS) technology evolved from MOSFET technology and is destined for the development of Very Large Scale Integration (VLSI) systems. Thanks to the properties of complementary MOS transistors, this planar technology (i.e. specifically adapted to surface transistors that we find on chips) has enabled the creation of low-cost and low energy circuits. This advantage has meant that this technology is recognized as the central technology behind the microelectronics industry (the bipolar transistor, which was referred to earlier, also exists in planar form and also has the advantage of being very fast; however, its sheer size and its large energy consumption limits its usage).

The underlying idea behind CMOS technology is to create pairs of complementary transistors (i.e. a P-type coupled transistor and an N-type coupled transistor). Each pair is able to create logic gates based on Boolean principles used in digital electronics.

16 Nanocomputers and Swarm Intelligence

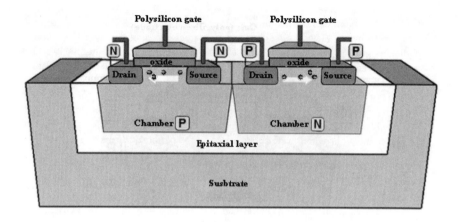

Figure 2.7. *A Complementary Metal Oxide Semi-conductor (CMOS) transistor*

Several types of CMOS technologies exist. We will only refer to the most common ones: N-type chamber CMOS technology, P-type chamber CMOS technology and double chamber CMOS technology. N-type chamber technology uses a P-type silicon substrate in which an N-type silicon chamber is formed. P-MOS transistors are then created in this N-type chamber and complementary N-MOS transistors are created in the P-type substrate. P-type chamber technology has a reverse effect: P-MOS transistors are created in the N-type substrate and N-MOS transistors are created in the P-type chamber. With regards to double chamber technology, it uses either the N or P-type silicon substrate on which P-type or N-type chambers are deposited.

2.2.3. *Half a century of industrial methods and processes*[1]

The production processes used for the development of semi-conductors and other electronic components have greatly evolved since the invention of the first transistor. They have given rise to the integrated circuit, followed by the high-density integrated circuit, and also to elements the size of a coin and to elements which are hundredths of nanometers in size. However, silicon technology has largely remained the same.

---

[1] The current section aims to shed some light on the industrial processes used in the production of semi-conductors. The reading of this section is not compulsory. If you choose not to read it, it will not affect the understanding of the rest of the book.

The Rise and Anticipated Decline of the Silicon Economy 17

> The first integrated circuit was developed in 1958 by Jack Kilby who worked for Texas Instruments. These circuits were tiny piles of rotating layers of semi-conductors and insulators which were capable of constructing, in the block of a circuit, a certain number of transistors. These transistors, which were interconnected into macroscopic circuits in the same block, allowed for the creation of memory devices, as well as arithmetic and logic units. This revolutionary concept concentrated on a maximum number of logic functions in an unbelievably reduced space. These functions reached the exterior of the unit through connections which were spread out over the periphery of the circuit.
>
> In order to create these chips, Texas Instruments developed a process of production by diffusion in the semi-conductor which became known as "planar" and which is still used today.
>
> In 1968, Robert Noyce and Andrew Grove, who were unable to convince Fairchild into focusing on the area of integrated circuits, founded their own company INTEL. INTEL is a contraction of INTegrated and ELectronics.

Silicon is obtained from sand and, more specifically, from very pure sand which is very rich in silica and which is relatively low in "pollutants" such as sodium. This silicon, which is still unsuitable for industrial use, is refined by using the process of purification called zone melting. The bar of silicon is introduced into a neutral atmosphere or vacuum oven and is subjected to a process of heating by induction or by laser, but only for a limited period which leads to the fusion of a very weak wafer of the bar of silicon. In these conditions the impurities of the almost solid zone spread to the liquid microzone which becomes richer in waste products, whereas the solid zone is purified. It is only necessary to slowly move the heating zone from one extreme of the bar to the other in order to transport most of the impurities to the unsuitable extreme which will be eliminated. By proceeding with successive scanning, we progressively succeed in creating a bar of silicon which is suitably pure.

The pure or intrinsic silicon which has been created will undergo two types of treatment:

– a gain in doping material in order to make extrinsic N- or P-type silicon, which are made up of a determined proportion of impurities with controlled electronic characteristics;

– a transformation of the bar (which at this stage is a tangle of silicon crystals piled up one on top of the other) into a suitable monocrystal.

This operation is carried out in a pulling oven. The silicon is melted in a neutral atmosphere, and a suitably positioned monocrystalline germ is brought in contact with the liquid and is slowly pulled towards the top of the oven and is simultaneously slowly rotated. The liquid is carried away by capillarity and is solidified by the spreading of the crystalline germ. This is what is known as the epitaxy process. Doping is obtained by introducing a predefined concentration of the additive into the liquid bath.

After recooling, the bar is cut into slices which are 100-300 μm thick. These slices are known as wafers. These wafers are almost exclusively manipulated by robots and are stored on airtight cassettes. Also at this stage, all the operations are carried out in an atmosphere which is completely free from dust (a clean room with critical zones which are composed of less than one particle with a size of more than 1 micron per cm^3).

We then move onto the development of active layers. Different processes co-exist: thermal diffusion, the implantation of ions and epitaxy.

The traditional process was based on thermal diffusion. The wafers are placed in a tube heated between 900 and 1,000°C according to the operating phase. The tube is also constantly scanned with a flow of neutral gas such as nitrogen or argon. This flow will favor the diffusion of the doping element in the first phase. This is followed by an oxidation phase in which oxygen and a weak proportion of steam is added. Finally, in the reheating phase (homogenization thermal treatment) only the neutral gas scans the tube. A variant of the diffusion process takes place in a sealed bulb. In this case, the doping element is placed in a cupel which is heated to a temperature greater than that of the silicon wafers. The boron (or arsenic) passes to the gaseous state and will condense at the surface of the silicon wafers, then just as in the previous case the boron (or arsenic) will spread to the inside of the material.

The process of thermal diffusion leads to a concentration of doping atoms which progressively decrease from the surface. However, PN junctions are established and lead to the creation of abrupt junctions. In other words, this creates a brutal passage from a zone which is made up of P-type doping elements to a zone which is made up of N-type doping elements. In order to try and homogenize the diffused zone, a process of thermal treatment is applied once the diffusion process has stopped. This is not always enough in order to create a very abrupt junction. Sometimes ionized atoms are directly implanted with the help of an ion implanter. It is possible to create perfectly localized zones where junctions are abrupt as well as zones with too many resistant properties.

A third process to create an active layer, stemming from thin layer technology, consists of slowly depositing a suitably doped silicon layer on the wafer of the opposite type. This operation is carried out either by the simultaneous evaporation in a vacuum of the silicon and the doping element which comes from two independent sources, or by using the vapor phase from SiH_4 and a component of the doping element. The silicon wafer is heated to a sufficiently high temperature enabling the atoms to take their correct positions, i.e. continuing the monocrystal without introducing any flaws at the interface (epitaxy process). This temperature remains, of course, under the maximum limit required for the thermal diffusion process to begin. Relatively thick abrupt junctions are created because the thickness of the epitaxed zone can be relatively large.

Several of these processes are combined in the creation of a complex component.

However, the creation of an integrated circuit reveals a more complex approach than simply creating suitably doped active layers. It is a question of implanting all of the primary components of the circuit onto the silicon wafer. The first step of this process consists of taking principles from the world of electronics and using them in the implantation process. We must define the respective locations for each component as well as their superficial geometry by taking into account doping requirements, the thermal dissipation of the functioning component, as well as the connections towards the exterior of the circuit and to all neighboring constraints in terms of variable electronic fields. Today, this type of operation relies on computer-aided design software which enables simulations and offers crucial support in the resolution of complex topology problems, which are related to this type of micro-urbanism. Furthermore, these computing tools are equipped with vast libraries of primary components and predefined modules, allowing for the easy industrialization of silicon.

Beyond the resolution of topology problems, this simulation stage also enables the optimization of scheduling operations in the operating range, i.e. identifying all production operations which are simultaneously carried out. For example, for a given P-type doping element (base NPN transistors, emitter PNP transistors known as substrates, emitter and collector PNP transistors known as laterals), base resistors and base pinch resistors will be created during the same diffusion operation.

The result of this implantation process (i.e. in the topology of a P-layer in the current example) is a large-scale drawing of the layer for a basic circuit. Traced with the aid of a computer, this black and white drawing will then be photographed and reduced in size and it will then be reproduced many times (like a sheet of postage stamps) on photographic film. This film, once developed, will serve as a projection matrix on a screen and a new photograph will be taken allowing for the second reduction in size of the drawing.

This procedure of photographic reduction will be repeated until a photo plate with the desired number of patterns is obtained. (A scale one photo plate is very fragile, so a replica in the form of a chrome mask is often preferred. To do this, a chrome layer is deposited on a glass plate on which the photo plate is recopied by engraving it onto the glass plate. The chrome sticks to the glass. The mask, which has also been created, can be manipulated and used many times.)

The silicon wafer is first oxidized at the surface at a depth of 1 µm in an oxidation oven (similar to that used in diffusion) in which the temperature is maintained at 900°C for several hours with the presence of oxygen and vapor. The silicon wafer is then coated with a uniform layer of photosensitive resin approximately 1 µm thick with the help of a high-speed turning device, enabling a uniform spreading of the resin with no microbubbles of air present.

> Photolithography consists of depositing a layer of photosensitive resin on a small plate of oxidized silicon. This resin is then irradiated by the use of a mask, which is made up of part of the pattern of the chip. Irradiation changes the solubility of the polymer, becoming either soluble or insoluble. The most soluble part is removed thanks to a developer (an aqueous solution or organic solvent).

After drying the resin, the wafer is placed into a stepper in which the photo mask is aligned opposite the wafer with the precision of a fraction of a micron. We then proceed with the insulation of the resin through the mask by using an intense flow of perfectly homogenous UV light which is perpendicular to the wafer. A solvent is used in order to dissolve the non-exposed zone of the resin. This process, which stems from graphic art lithography, is known by the term photolithography.

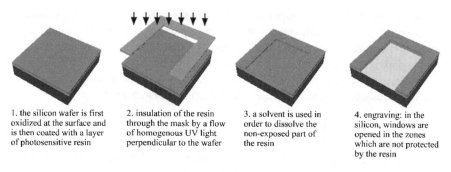

1. the silicon wafer is first oxidized at the surface and is then coated with a layer of photosensitive resin

2. insulation of the resin through the mask by a flow of homogenous UV light perpendicular to the wafer

3. a solvent is used in order to dissolve the non-exposed part of the resin

4. engraving: in the silicon, windows are opened in the zones which are not protected by the resin

Figure 2.8. *The process of photolithography and the engraving of silicon technology*

With the arrival of high-density integrated circuits, i.e. chips which have several hundred million transistors (such as the Pentium 4 HT with its 167 million transistors, the HP PA-8 800 which has more than 300 million chips or the new Itanium² and its two core chip Montecito which has 1.7 billion transistors), the development of the mask and its precise positioning have become extremely critical. This is why it is often preferable to use another process which consists of directly engraving the pattern on a scale of one in the resin by using an electronic beam which is managed by a computer. This basic pattern is repeated on the wafer after the displacement of the pattern by a step, then by a line until the useful area of the wafer is covered. This operation is carried out in a vacuum.

Figure 2.9. *A silicon wafer containing multiple circuits*

The major interest in such a procedure, in addition to the removal of the operations of mask-making, resides in the finesse of the electronic beam which makes it possible to obtain resolutions smaller than 10 nanometers, which the high integration of components demands.

The next operation is engraving. In the silicon, windows are opened in zones that are not protected by the resin. Several techniques are possible. The simplest technique consists of chemically dissolving the non-protected silica with the help of a fluorhydric acid, then removing the resin by a process known as mechanical and/or chemical stripping.

The wafer is now ready to go through a new diffusion process.

After multiple operations of creating active layers, superficial oxidation protects the entire wafer. Windows linked to areas of contact are re-opened. An aluminum, sometimes gold, and more recently copper layer (2 to 3 μm thick) is deposited on the wafer, which is then engraved following a similar procedure to that of litho engraving mentioned previously.

Certain circuits can be faulty, either because they are located on the edge of the wafer i.e. they are incomplete, or because some dust has interfered with the masking process. Before being cut, these faulty circuits are identified and located by the use of a simple test involving a drop of magnetic ink, which enables them to be eliminated after cutting by simply passing a magnet through the circuit.

Test zones, which are made up of different patterns, are used to globally verify if each successive masking operation has been suitably carried out. These test zones are developed on the wafer. In effect, it is not possible to test the integrated circuit. The primary elements of the integrated circuit are too small to be reached by a test probe without damaging them.

After all the operations used in the creation of active layers have finished, the silicon wafer is cut. Each circuit is mounted onto a support. After soldering the connections of the chip to the exit port, the component is placed in a capsule.

There are two common techniques which are used in order to assure the connections with the exterior: ultrasound and thermocompression. By using aluminum wire with a diameter of 25-50 microns, the ultrasound technique enables the aluminum layers to be connected to the external prongs of the component. It is not really a question of soldering but rather sintering. The wire, which is applied to the connecting zone, is strengthened without the entire unit having to undergo the process of fusion. The wire, in the connecting zone, is strengthened with the help of a high frequency vibrating tool. Thermocompression is preferable when dealing with the internal bonds on a hybrid circuit between layers of gold (or copper). In this case, a micro-blowlamp assures the fusion of the extremity of a gold wire, which is then immediately applied with significant pressure to the location to be connected.

> In 1969, the Japanese company Busicom developed extremely powerful programmable computers. The company sub-contracted the production of a set of 12 integrated circuits to Intel. Marcian Hoff and Federico Faggin considered the convenience of reducing the number of circuits by concentrating a large number of logic functions on each one of them. Instead of the initial forecast 12 circuits, they developed a computer which would function on a chipset of only four circuits (MCS-4), amongst which there would be a chip dedicated to computation. The 4004 would become the first microprocessor in history.
>
> The chipset idea is still present in modern computers. Although today there are only two integrated circuits which complete the functions of modern microprocessors: Northbridge (which is in charge of controlling exchanges between the processor and the live memory) and Southbridge (which manages communication between the input and output peripheries), chipset coordinates the data exchange between the different components of the computer.

2.3. The fatality of Moore's Law: "the wall"

These integrated circuits have seen their power continually increase for more than 40 years now. We talk about the density of integration in order to express the quantity of transistors that can be put on the same chip.

In 1995, the powerful Alpha chip by Compaq (a 64-bit processor, which was a technological spearhead at the time, even if the circuit's market share remained average) was made up of approximately nine million transistors. Six years later, a state of the art microprocessor would have more than 40 million transistors. In 2015, it is believed that these processors will contain more than 15 billion transistors.

After predicting an SRAM memory using 65 nm technology in the third trimester of 2004, TSMC (Taiwan Semi-conductor Manufacturing Corporation) launched the first low-power chips (i.e. memory circuits) engraved with an engraving of 0.065 microns at the end of 2005. The first 65 nm high-power chips (i.e. processors) are forecast for release in the first half of 2006.

By the end of 2007, Advance Micro Devices (AMD) feels it can change the course of all 65 nm production in its new factory, Fab 36. One fabrication plant will stop using engravings of 65 nm and will change to 45 nm starting from mid-2008 onwards.

Things are becoming smaller each year. If this continues, in theory, in less than 10 years computers will be created where each molecule will be placed at its own place, i.e. we will have completely entered the era of molecular scale production.

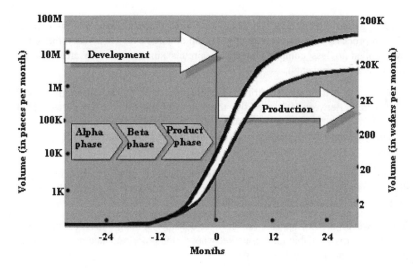

Figure 2.10. *The acceleration of "time to market". For some years now the cycle of technological development has had a periodicity of two years, in comparison to three as used to be the case. Furthermore, the scale factor of miniaturization which was typically 0.7 between two generations (for example, going from 130 to 100) is no longer valid. In order to respond to market demand more often and in a quicker manner, "half-generation" intermediaries (180/150/100) have now been introduced*

This law of miniaturization that governs the silicon industry is known as Moore's Law. It states that every 18 months, the size of transistors will be halved and, as a consequence, the processing power of microprocessors will be doubled ("the number of integrated transistors on a silicon plate of a given size is multiplied by four every three years" Moore, 1965).

After having closely followed this prediction for more than three decades, the process has, however, accelerated over the last few years, reaching a doubling in the performance of processing power every 12 months, instead of 18 as was initially forecast.

> In 1965, Gordon Earle Moore, co-founder of the Intel group, was working for Fairchild. Whilst preparing for a presentation published in *Electronics Magazine*, he made an interesting observation regarding memory chips. State of the art circuits available on the market doubled their working capacity approximately every 18 months.
>
> In predicting this tendency he observed an exponential evolution in memory capacity as well as in the processing power of microprocessor chips (remember that in 1965 the most powerful circuit had 64 transistors).
>
> This exponential growth was christened Moore's Law or, more precisely, Moore's first law.
>
> In 1980, Moore expressed a second law, this time according to which the number of microprocessor transistors on a silicon chip doubled every two years (the initial law applied to less complex integrated circuits since they were created from independent components).
>
> Although this is a simple, empirical law, this prediction was astonishingly proved to be correct. Between 1971 and 2001 the density of transistors effectively doubled every 1.96 years. This tendency still holds true today. It should, in principle, remain valid until 2015 when the process will no longer be able to avoid quantum effects.
>
> Just as an anecdote, in 2004 the theory experienced a slight deceleration due to difficulties of thermal dissipation, hindering a rise in density even though the size of the components continued to get smaller.
>
> Still, however, the law appears robust if we consider, among others, the new two-core processors available on the market whose global performance doubles whilst maintaining an unchanged clock speed.

The cost of technology enabling the creation of chips with more and more transistors is increasing by breathtaking proportions. Another empirical law from Silicon Valley, Rock's Law, states that the cost of producing a chip doubles every four years because the production process, photolithography (see section 2.2.3), which has been used for over 40 years, is reaching the end of its physical limits.

In 2004, Intel announced that the company was planning to invest more than 2 billion dollars in its factory Fab 12 in Arizona for the production of chips from 300 mm wafers. This new dimension progressively replaced 200 mm wafers whose technology expired by the end of 2005. With a diameter larger than one-third of the 200 mm wafers, the 300 mm wafers create a 225% gain in area in comparison to their 200 mm predecessors. In other words, with this fifth factory using 300 mm technology, the Intel group has a production capacity equivalent to that of 10 factories using 200 mm wafers. The more economical 300 mm wafers are a prerequisite for the industrial use of 65 nanometer technology, which was used at the end of 2005. The engraving process is currently being developed in Hillsboro, Oregon and is the latest of the Intel group's development units. This development unit known as D1D has received investment worth $2 billion.

The circuits are engraved on very thin silicon wafers by focusing a ray of light through a mask which contains half the patterns of the circuits and transistors. The resolution (or thickness of the transistor mask) is confined to half the wavelength of the light's radiation.

In order to produce even smaller transistors, radiation with shorter wavelengths is used and the race to miniaturization has progressively led to the creation of harder X-rays (such as UV radiation, X-rays, X-UV, etc.). However, in this range of wavelengths it becomes difficult, indeed impossible, to effectively concentrate the rays.

In the mid-1990s, it was thought that it was not possible to industrially produce transistors with a diameter of less than 400 atoms (100 nm) with such a process.

Today it is considered possible to push back the critical dimensions of CMOS transistors to approximately 20 nm. Demonstrations have already taken place in the laboratory (see section 4.2.1).

However, this final dimension will lead to both a physical and industrial limit for this technology. In the silicon industry this limit is referred to as "the wall".

26 Nanocomputers and Swarm Intelligence

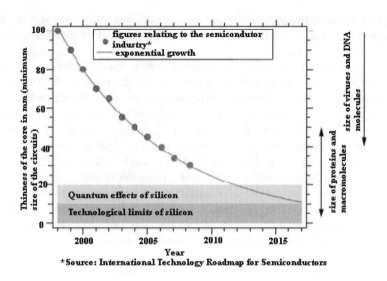

Figure 2.11. *The "wall" or physical limits of traditional silicon technology*

In 1999, the so-called ultimate CMOS transistor developed at the Laboratory for Economics and Information Technology in Grenoble, France, tested the limits of the principles of the MOSFET transistor with a diameter of 18 nm (approximately 70 atoms placed side by side). This was almost 10 times smaller than the smallest industrial transistor in 2003 (130 nm in 2003, 90 nm in 2004 and 65 nm in 2005). It enabled the theoretical integration of seven billion junctions on a €1 coin. However, the CMOS transistor, which was created in 1999, was not a simple research experiment to study how CMOS technology functions, but rather a demonstration of how this technology functions now that we ourselves are getting ever closer to working on a molecular scale. Today it would be impossible to master the coordinated assembly of a large number of these transistors on a circuit and it would also be impossible to create this on an industrial level.

2.3.1. *The disaggregation of the microelectronics industry value chain*

Today, approximately 200 companies worldwide work in the equipment and raw materials sector for the semi-conductor industry. These sectors supply the companies who come up with, produce and distribute semi-conductor components and other devices linked to semi-conductors.

During the pioneering phase of this industry the producers of integrated circuits created their own production equipment and integrated the development of semifinished materials into their production process. The manufacturers of semi-conductors took care of everything, from the conception to the delivery of the integrated circuit to the component manufacturer. The latter ensured the support phase of the cycle, from the chip to the final customer.

The customer was either a client, as in the case when purchasing and constructing a car, or a consumer (for example, television, audio or other consumer products on the market). With the growth and maturity of the market, the chain of suppliers and components manufacturers has become progressively specialized.

The responsibility for the development of new methods of production and production lines lies increasingly with the components manufacturer such as Applied Materials Corp, Nixon and KLA-Tencor. Over the last six to eight years there has been a real disintegration of the chain. This can be resolved by having more people in the chain with only one job or being involved in one specific activity which they can master at the top level on a global scale to cope with global competition.

This economic model which is based on hyperspecialization is not a new concept; the car industry has been using this concept for 50 years. It creates real industrial strength, almost hegemony of companies which possess the industrial excellence of a specific process or procedure, for example, the fabrication plant of the Taiwanese company Taiwan Semi-conductor Manufacturing Co. (TSMC). Some exceptions still exist: it so happens that large producers of components (such as IBM in this example) directly take part in the development of new processes and procedures when they believe that they can remove a competitive advantage. However, these cases are very specific and do not occur all that often.

The maturity of an industry generally leads to two consequences: on the one hand, the specialization and disaggregating of activities, and on the other hand the emergence of standardization which comes naturally with this part of the chain. This standardization affects the structure of microprocessors as well as the manufacturing processes, logistics and the operating software.

28   Nanocomputers and Swarm Intelligence

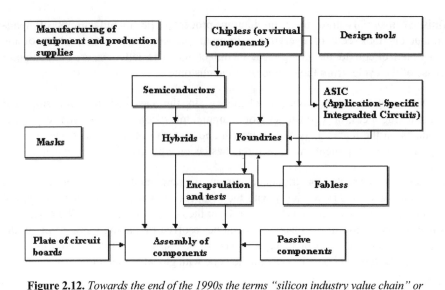

Figure 2.12. *Towards the end of the 1990s the terms "silicon industry value chain" or "microelectronic clusters" were used when referring to corporate networks which focused on the silicon industry. The aim was to take advantage of strategic alliances which were dictated by the costs of research and development and investment by using prohibited means of production. In effect, these costs had reached new levels which were no longer accessible by integrating all, or a prohibitive number, of functions (today a state of the art factory represents more than 400 times the investment that was needed to be agreed upon in 1974). Furthermore, when the market for chips experienced periods of low demand (periodic fluctuations), it was better to be extremely specialized and to concentrate as much as possible on a perfectly targeted sector. Generalist manufacturers of integrated circuits initially gave way to "fabless and foundries". Original and distribution suppliers of circuits trusted in the large foundries for the production of the chips which the former had developed. Foundries then concentrated on the production of the chips and could agree to the investments that were imposed by the ever increasing process of miniaturization. The two large fabrication plants UMC and TSMC announced their usage rates as greater than 90% for state of the art engravings (130 and 90 nm), compared to 40% with traditional manufacturers of circuits. This information was recorded in 2001. Are there also more traditional manufacturers of circuits who adopt the fabless concept and sell their internal fabrication plants? A third type of player has appeared more recently, the idea of chipless operators. These new chipless operators specialize in the design of logical building blocks that are sold in the form of intellectual property in return for licenses and royalties. In the field of component assembly the tendency has been to migrate to regions where there is a low-cost workforce for mass production and to develop partnerships with principal clients in order to optimize design and prototyping techniques*

This tendency towards the functional specialization in the semi-conductor industry has no reason to slow down in the years to come. The average cost of a fabrication plant commonly known as Fab grew by nearly 15% a year between 1983 and 1999 (source, Dataquest) going from $200 million to $1.8 billion (180 nm technology). A factory producing 300 mm wafers engraved by 90 nm technology today costs up to $4 billion. Few operators throughout the world are able to make similar investments. Intel and IBM are part of this elite group, while AMD has left the group. Rock's Law and its impact on the access price to new generations of semi-conductor producers has introduced a new type of wall. This wall is not strictly technological but more economical. Even though the unitary cost of a transistor has continually decreased since its creation, the cost of factories able to produce them has, however, continually increased which slowly disqualifies potential candidates on one criterion: being able to afford the production tools.

In order to make a profit it is necessary to concentrate a lot of work on these factory units. If certain companies resorted to strategic alliances to stay in competition, such as Motorola who mutualized its production methods with Philips and STMicroelectronics, then a good number of average size companies would consider that it is no longer economically viable to operate their own production and would move towards a model "without factories" in which the production of their components is sub-contracted elsewhere. This migration established the advent of large silicon fabrication plants such as TSMC, which was mentioned previously, and United Microelectronics Co. (UMC) who are exclusively in charge of the production of chips.

2.3.2. *The ITRS (International Roadmap for Semi-conductors) – a race to controlled miniaturization*

With the progressive externalization of production tools to the suppliers of specialized equipment, the need has arisen for a clear roadmap to anticipate the evolution of the market and to plan and control the technological needs of subjacent production. For several years, the Semi-conductor Industry Association (SIA) has given this responsibility of coordination to the USA, which led to the creation of an American style roadmap, the National Technology Roadmap for Semi-conductors (NTRS). In 1998, the SIA became closer to its European, Japanese, Korean and Taiwanese counterparts by creating the first global roadmap: The International Technology Roadmap for Semi-conductors (ITRS). This international group had 936 companies which were affiliated to working groups for its 2003 edition.

This forward planning organization establishes a vision for the future from year to year, which should include state of the art silicon technology over a period of five years. On a more long term scale (from 6 to 15 years), the projections are elaborated over three year periods. For each field studied, the ITRS identifies the technological difficulties and challenges that the industry will have to overcome for the period in question in order to preserve the envisaged progress for the industry.

30 Nanocomputers and Swarm Intelligence

Year of first product delivery	2003	2005	2007	2009	2012	2015	2018
Characteristic size (MPU/AISC uncontacted Poly)	107 nm	80 nm	65 nm	50 nm	35 nm	25 nm	18 nm
DRAM capacity	4GB/ 4.29GB	8GB/ 8.59GB	16GB/ 17.18GB	32GB/ 34.36GB	32GB/ 34.36GB	64GB/ 68.72GB	128GB/ 137.4GB
Number of transistors/microprocessor chips	180 million/ 439 million	285 million/ 697 million	453 million/ 1,106 million	719 million/ 1,756 million	2,781 million/ 3,511 million	5,687 million/ 7,022 million	9,816 million/ 14,045 million
Number of metallics layers of cabling (min/max)	9/13	11/15	11/15	12/16	12/16	13/17	13/17
Clock speed (GHz) of microprocessor "on chip"	20976	5.204	9.285	12.369	20.065	33.403	53.207
Diameter of wafer	300 mm	300 mm	300 mm	300 mm	450 mm	450 mm	450 mm

Source: International Technology Roadmap for Semiconductors 2003

Table 2.1. *The ITRS (International Roadmap for semi-conductors): a 15 year plan*

State of the art microprocessors at the end of the 20th century had a maximum of 25 million transistors (28 million for the Pentium III processor in October 1999) whose characteristic dimensions come close to 250 nm. In 2003, engraved with a core of 130 nanometers, the Pentium 4 HT had 167 million transistors. In 2004 the size of cores available on the market reached 90 nm and became 65 nm in 2005. This goes beyond the 1999 predictions made by the ITRS who predicted that the reference microprocessor in 2005 would have only 190 million transistors and an engraving of 100 nm. The ITRS believes that the characteristic dimension of transistors will reach 50 nm between now and 2011, and 35 nm between now and 2014. It would therefore be possible that by 2014 microprocessors based on silicon technology will have become faster and that electronic components made from silicon will have reached their smallest dimensions possible.

> Intel's strategy has been taken from the Ford model at the beginning of the century: a lot of factories which produce a lot and which are the leaders in engraving techniques. With each change in the generation of engraving, one of the generations is dedicated to the pilot development of new technology. When the transition phase has ended, this new technology is introduced into other factories (a process known as roll-out) with a very short deadline so as not to harm production capacity and to benefit from the technological advantage of a commercial life cycle for as long as possible.

The American inventor Ray Kurzweil (the principal developer of the first omni-font optical character recognition) confirmed that information technology is moving onto the second half of the chessboard, referring to the story of the Emperor of China who offered to give the Prince anything he wanted by granting all his wishes. All the Prince asked for was rice.

"I would like one grain of rice on the first square of the chessboard, two grains of rice on the second, four grains of rice on the third square and so on, doubling each time all the way to the last square." The emperor agreed, thinking this would add up to a couple of bags of rice.

What a poor estimation. Quietly progressing through the multiplications, the number of grains of rice per square breaks the two billion barrier halfway through the chessboard (2^{31} on the 32nd square, which is 2.147 billion grains of rice). The final square contained nine billion billion grains of rice, in other words, 500 billion m^3, which is enough to cover the Earth's surface with a layer more than 10 cm thick (which is 2^{63} or $9.22337024 \times 10^{18}$ grains of rice on the 64th square).

This law, known as geometric progression, is very similar to the one which governs the evolution of the number of transistors on silicon chips.

In this way, we are almost halfway through the chessboard today. The first microprocessor produced by Intel in 1971 contained 2,300 transistors (the 11th square in our chessboard). Since then, a new chip has been created every 18 months which contains twice the number of transistors than the previous chip. In 2003 a chip contained 500 million transistors (the 19th square). Planned for 2007 in the 2003 Roadmap, the first chip with one billion transistors was introduced in 2005 with Intel's Montecito processor (the 30th square). It must be made clear that in terms of the number of transistors integrated on silicon chips, the one billion mark had already been reached in 2003 with the release of 1 GB memory chips, which had exactly 1,073,741,824 transistors, each transistor being the equivalent of 1 bit.

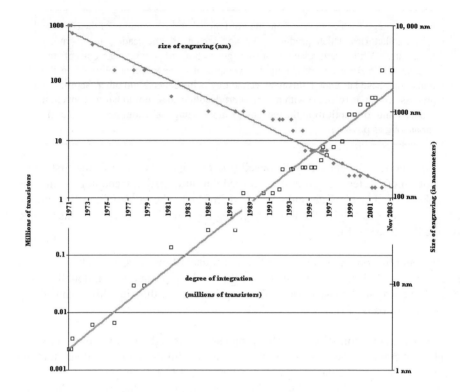

Figure 2.13. *The exponential evolution of processors. The degree of integration and finesse of engraving*

However, is it opportune to consider that the limit of 15 billion transistors on future chips by 2015 will constitute an unavoidable change in progress? Aren't there any solutions which would avoid this? In fact, the inflation of electronic power and the facility with which these resources have been growing for four decades has led to an enormous waste of resources such as memory and the number of machine micro-instructions needed in order to carry out a basic function. The execution of such a simple program as comparing two objects in order to give an "identical" or "different" result today needs 50 to 500 times more system memory than was needed 25 years ago because programming uses languages and interfaces that resemble more and more the natural language of the user. Throughout time, information systems have piled up layers of increasingly abstract software. Suppliers and service providers no longer have the ability to master its global complexity. Certain seemingly simple processing tools (selected from a large database) often using more than 10 pieces of different software have led to problems of which the most complicated require hundreds of people working days to resolve. The different software which makes up these processing tools is used in areas such as the microprogramming of processors and the management of parallel systems up to the highest application layers of Enterprise Resource Planning. The global integration of such software into the company's system, and even the meta-organization framework representing corporate networks with their customers and suppliers, also leads to the problems previously mentioned. This complexity has created more complexity. This unavoidable disorder has been caused by an increase in the number of methods used. It is also a shift in progress which provides a potential threshold level for silicon technology.

2.3.3. Will a slowdown in the economy overturn established models?

The hopes founded in the ambitious economic models that were developed around the Internet and wireless telecommunications led to a period of marked economic growth during the second half of the 1990s. This was a period of growth which was brought about mainly by the telecommunications, component manufacturing companies for information system and networking sectors, but also by the rapid expansion of Internet companies. This growth was very beneficial to the development of the semi-conductor industry and to the acceleration of the miniaturization process which exceeded empirical predictions as established in Moore's Law.

However, this excess growth started to slow down in 2001. Growth, in fact, decreased dramatically. Massive job losses were recorded. These job losses were not limited to only new start-ups, or to traditional companies which strongly invested in e-commerce (click and mortar). This decline extended to all sectors of the computing industry on an international scale. The outbidding of the dotcom business, extremely high costs for purchasing user licenses for future generation wireless networks (2.5 and 3G, which were all part of this boom period and at the same time were economically unstable) strongly shook this sector of the economy. The explosion of the stock market bubble started in April 2000 in America, then, with the tragic events of September 11, 2001, the first signs of recession appeared. Industrial investment froze and, as a result, the main aim of companies was to limit themselves to reducing costs, thus jobs became more and more under threat and the first repercussions on consumer confidence were experienced.

Suppose that similar circumstances were to bring a fatal blow to the concept of miniaturization, which represents the reality of Moore's Law: then for a period of five years the industry would have to accept overwhelmingly large projects as well as massive investments often focused on information technology and an excessive debt paralyzing all advanced technological initiatives. For the past four years, third generation (3G) telephones and their succession of promises and very innovative applications do not seem to reflect the same interests (although analysts predict a continuing boom similar to that of the famous "killing applications", which were the PC or first generation digital mobile phone of their time).

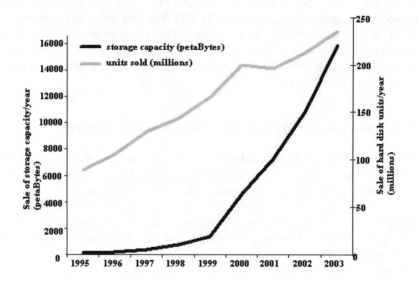

Figure 2.14. *Storage capacity and number of hard disks. Stable sales even in periods of slow economic growth*

Logic would have it that this slowdown in economic growth would directly impact performance in the field of electronics. Such a risk in terms of economic growth unconditionally leads to a change in the race to the infinitesimal. Its progression still continues in accordance with the predictions of Moore's Law. Digital technology continues its global penetration. Less than 20 years after its commercial launch, a quarter of the world's population possesses a cell phone which is a rate of penetration yet to be observed in technological history.

The access, exchange and storage of information have become a priority for all social classes in every country of the world. The dematerialization of cheap electronics has been the key factor behind this new era in the history of humanity. Whether it exists in physical format (paper, film, magnetic or optic) or whether it is telecommunicated (by telephone, radio, television or the Internet) more than 90% of all of today's information is stored on magnetic support of which the hard disk plays a key role (nearly two-thirds of information produced worldwide in 2002 was saved on this type of disk which has seen its installed base grow by more than 85% since 1999 and whose cost continues to fall: $1 per gigabyte, in other words $1 per billion characters at the beginning of 2003 (source: SiliconValley.com)).

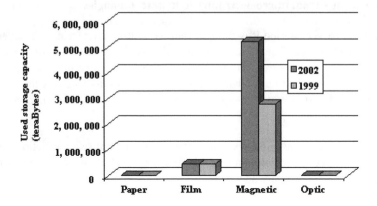

Figure 2.15. *More than 90% of all information generated is stored on magnetic support (mainly the hard disk)*

Microprocessors and computers are generally used for saving, reading, filing and processing digital information. The one billion users of the Internet (since 2006), new gadgets and applications in the mobile phone industry, infiltration from audio and video worlds and the emergence of new digital platforms such as GPS, embarked systems, PDAs and video games consoles, have continued to boost the growth of the silicon industry during these periods of stagnant or slow economic growth.

Of course, the tendency of miniaturization as expressed in Moore's Law is independent of the industrial margins of microelectronics whose 40 year history connects periods of strong unitary margins with discrete volumes, with mass market phases where larger volumes make up for lower unitary profits.

It is generally during these periods of democratization of advanced technology when large steps in the delocalization of an industry are observed. We remember the radio receptor and television industries whose assembly factories had been extensively relocated to the leading Asian countries such as Taiwan, South Korea, Hong Kong and Singapore, as well as to newly industrialized countries such as Brazil. The industry of entry-level and middle of the range televisions was no longer considered as a "high technology" for the USA, Europe and Japan who all started to concentrate their technological know-how on emerging digital technologies.

2.4. Beyond silicon – from microelectronics to nanotechnologies

Over the next 15 year period it is thought that we will be witnessing the advent of the ultimate technology. The most optimistic idea is that this technology could respond to the majority of our material needs. This utopia will come into force when production will have reached its fundamental limits and when industrial technology will be able to rebuild the world, atom by atom. This is what is known as nanotechnology.

If you imagine the future in 30 years as a vision taken from science fiction you might be wrong. However, if you imagine this future by excluding all references to science fiction you will definitely be wrong.

Predicting the future of technologies was one of the biggest obsessions in the 1960s, a decade in which dozens of Institutes of Futurology were created throughout the world. Although certain projects have only existed in science fiction novels, the real world has often exceeded expectations that at the time were considered as utopian.

Nanotechnology, for the most conservative scientists and engineers, seems to be an idea taken from science fiction. However, in the 1930s the thought of landing on the moon also seemed like an idea from science fiction. Yet, for the people concerned they knew that it was theoretically possible to land on the moon because they had the competence and the know-how as far as fuel chemistry, thermodynamics, mechanics, etc. were concerned. Today, we know that it is possible to do the same thing with regards to the manipulation of atoms. In understanding long term applications of nanotechnology, but also trying to attempt to explain where each stage will lead to, we can see that it is nothing but another chapter in the history of science and technology which is being written.

The concept of artificial nanomachines created by man has for a very long time been considered by the scientific community with reservation, and even more so by industrial investors.

The first creations of these nanomachines, although confined to large industrial laboratories and universities, have for some years now shown real opportunities. These hopes have catalyzed certain public initiatives from volunteering states such as the USA, Japan and finally Germany to name only some of the leading countries. Of course, the stakes are so high today that all member states of the Organization of Economic Co-operation and Development (OECD) will take part in any initiative in this field of science. It is a risky business but the stakes are so economically and socially high that these states must stay in contention.

How is nanotechnology so radically different from traditional technology and how will nanotechnology appear as a rupture in the continuing race to the infinitesimal which is taking place today?

All materials are made up of atoms. The properties of materials depend on atoms and their arrangement within the material. Transforming lead into gold is scientifically impossible. However, transforming carbon into diamond from a "simple" rearrangement of carbon atoms, from an atomic structure that exists in nature to another equally existing structure is, in theory, technically possible.

Since the beginning of time, macroscopic processes have been used in order to produce a machine or any manufactured object. Even if dimensions are becoming continuously smaller, macroscopic processes are still used. The materials are extracted, modified in physical and chemical terms, deformed by removing matter, assembled, joined and stuck to large-scale aggregates of matter compared to the size of the atoms.

The basics of production have not really changed over time. A process of extracting raw materials is followed by a process of formatting by either removing the matter or by deformation. Large civilizations were created thanks to the history of cut and polished stone, followed by more complex manufactured products. Approximately 200 years ago humanity entered the industrial era with the steam engine and railways. Then, in the second half of the 20th century, it reached what we can now call modernity with, among others, microelectronics and the miniaturization of technologies. This technology is still based on the same fundamental ideas as those used back at the beginning of history: from the flint to X-ray photolithography of the latest generation microprocessors, it is a question of working the matter by sculpting and deforming it. We start with a piece of material to work with a product and remove the waste. We start with the largest in order to get to the smallest. We start from the exterior to get closer to the interior. This is an approach known as top-down,

However, nanotechnology relies on the reverse process: it consists of starting with the smallest in order to get to the largest. It starts from the interior (the atoms) towards the exterior (machines and manufactured products). This is why we call this process bottom-up. Nanotechnology is therefore a discipline which aims at studying, manipulating and creating groups of atoms from the bottom to the top. Initially, the term nanotechnology did not differ from microtechnology by degree of integration, but by the development process: from the bottom to the top, atom by atom as opposed to from the top to the bottom by progressively reducing the size of the production process by removing matter and through deformation. This distinction is fundamental even if today all industrial processes with a size ranging from a few nanometers to 100 nanometers are listed as nanotechnology. Techniques derived from the current process of photolithography have led to a certain restriction in the understanding of nanotechnology. It is this type of process which is, for example, the origin of the famous nanoguitar created by electron beam engraving. This is a musical instrument which is 10 nm long with strings and a 50 nm diameter, and which vibrates at a frequency of 10 MHz. It is a guitar the size of a bacterium. These technologies are, in fact, the result of a continuing process of miniaturization of microelectronic technologies and which are no longer part of the rupture caused by molecular nanosciences.

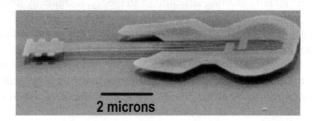

Figure 2.16. *An image of the world's smallest guitar made from silicon monocrystal created by researchers at Cornell University, New York. This is the only useful illustrative example available of science on a nano scale. This guitar, a copy of the famous Fender Stratocaster, measures only 10 microns which is the size of a red-blood corpuscle or 1/20 the thickness of a human hair. The strings have a thickness of 50 nm which is approximately 100 atoms placed side by side. If the guitar could produce sound, it would vibrate at a frequency of 10 MHz (source, Cornell University)*

The base for nanotechnology and its ambitious applications was formulated on 29 December 1959 by Professor Feynmann during his famous conference at the American Physical Society in the California Institute of Technology. In a legendary speech which would see the arrival of nanotechnology, Richard Feynmann (Nobel Prize laureate) stated that the laws of physics would not rise up against the possibility of moving one atom after the other: the laws of physics allow for the manipulation and controlled positioning of individual atoms and molecules, one by one, etc. This, in principle, can be accomplished but still has not been carried out in practice because we, as humans, are still too big. He is clearly remembered for illustrating his theory by the following example, which seemed rather avant-garde at the time: "Why can't we write the entire 24 volumes of the Encyclopedia Britannica on the head of a pin?"

In 1989, this prediction of extreme miniaturization (by a factor of 25,000 using the example of a pinhead) of engraving a text on an atomic scale was carried out in the laboratory. Donald Eigler and E. Schweitzer drew the initials of IBM by placing xenon atoms on a nickel surface with the help of a scanning probe microscope (SPM).

The Rise and Anticipated Decline of the Silicon Economy 39

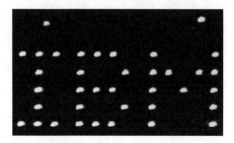

Figure 2.17. *In 1990, Donald Eigler and Erhard Schweitzer succeeded in engraving the initials of IBM by individually depositing 35 xenon atoms on a nickel plate with the help of a scanning probe microscope (source, IBM)*

We can imagine making all sorts of really amazing things with the knowledge that we currently have. Today we are no longer amazed with the abilities and capacity of microcomputers and we consider it as easy and normal to be able to manipulate the windows on a computer screen. We can dream of introducing significant intelligence capabilities at the heart of the primary components of objects, such as in molecules, so that the molecule can locate our presence and carry out appropriate action.

What about large-scale feasibility? All living cells use electronics the size of a molecule. Nature has produced its own nanomachines 100 times smaller than those currently used in silicon chips. This opens up a huge perspective for potential large-scale industrial applications.

Chapter 3

Rebuilding the World Atom by Atom

3.1. Manipulation on an atomic scale – the scanning tunneling microscope

"I think that in the middle of the 21st century whenever you see an airplane it will not be made from aluminum or from a graphite composite structure, but rather from a diamond structure. When you see a small, light, solid car which is safe and economic on fuel, it too will be largely made from diamond or from a material similar to diamond." These are the iconoclastic words used by K. Eric Drexler during an interview in 1994 in order to express the origin of "nanotechnologies". "In the same way that we experienced a Stone Age, a Bronze Age, an Iron Age and a Steel Age, I believe that we are going towards a Diamond Age or an age of materials inspired by diamond."

Today, diamond is a rare material and one of the most expensive there is, however, its element is neither rare nor expensive. For example, gold and platinum are made up of rare elements and this is why these two metals would conserve their rarity even if the arrangement of their atoms was altered. What makes a diamond so valuable is simply the arrangement of its carbon atoms. If you take a simple piece of coal and reorganize its atoms you would, at the end, get one of the most expensive materials available today on earth.

The diamond is the dream material for industrialists. It is unbelievably resistant, light and has excellent optic and semi-conducting properties. It possesses properties lacking in most other materials, and is the universal response to the majority of problems that other materials have.

Let us consider any object taken from our everyday lives, for example, a bicycle. It is, regardless of its technology, relatively heavy, rather fragile and its rigidity often leads to average performance.

If we could manufacture the bicycle according to a diamond structure on a molecular scale it would be of an unequalled lightness, extraordinarily rigid and perfectly solid. The diamond is much more resistant than steel and is 50 times harder.

Molecular manufacturing will allow for the manipulation of essential elements which make up the building blocks of matter (i.e. atoms and molecules) in the same way that the bits and bytes of a computer are manipulated. Atoms and molecules will be rearranged in order to make new structures in a quick and safe manner.

By definition, the aim of technology is to produce objects by manipulating the building blocks of matter.

It must be said that today we are still unfortunately incapable of obtaining such tiny pieces and putting them exactly where we want them.

In fact, we only know how to push them in "large packets". This is why a fundamental revolution with the most far-reaching implications will probably take place and it will change our way of producing absolutely everything.

The predictions of Eric Drexler regarding the positional control of atoms have undoubtedly become a reality.

In 1981 the German scientist Gerd Binning and the Swiss scientist Heiprich Roher from the IBM laboratory in Zurich developed an instrument which gave the first glimpse of the atomic world. It was an instrument which would open the door to the experimental reality of nanotechnology, the scanning tunneling microscope (STM), for which they won the Nobel Prize in 1986.

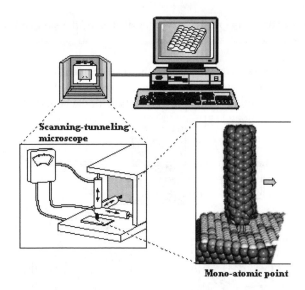

Figure 3.1. *The scanning tunneling microscope (STM) (source: "Unbounding the Future" – Eric Drexler and Chris Peterson)*

In this device, a probe whose extremity is mono-atomic scans the conducting or semi-conducting surface to be analyzed. By detecting very weak electric currents introduced by the atoms from the sample line by line, an image of the atomic topography of the surface is reconstructed.

However, during the first attempts the temperature made the particles move permanently and the first negatives were blurred. In order to improve the quality of the images scientists therefore continued to carry out a recooling process at a very low temperature of the device in order to limit thermal agitation on an atomic level. Yet the result was still not satisfactory because some free atoms "were jumping" onto the extremity of the microscope's probe to which they remained attached. Scientists then discovered that by inverting the polarity of the probe they could make the atoms fall anywhere that they wanted on the surface of the sample. The age of atomic manipulation had begun.

In 1990, the STM was used for the first time in order to produce complex structures on a nanometric scale.

In the STM the signal is based on the only electrical field between the probe and the surface atoms without any mechanical contact between them. In the case of the atomic force microscope (AFM), there is mechanical contact between the probe and the surface atoms. When the probe scans the surface of the sample horizontally, the atoms are optically detected.

Since 1990, the AFM has allowed for the displacement of atoms one by one in order to produce complex structures on a nanometric scale. Scientists have now put forward different methods of positional assembly with the help of this device.

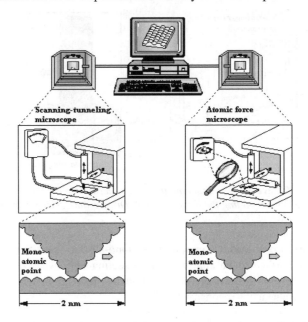

Figure 3.2. *The STM and the atomic force microscope (AFM). (Source: "Unbounding the future" – Eric Drexler and Chris Peterson)*

Traditional methods of studying and manipulating nanometric objects consist of working "in parallel" on large populations of objects. It is still like this today and it is only in this way that it is possible to produce an artificial macroscopic object. With the arrival of local probe measuring devices (i.e. STM and AFM microscopes) on the one hand, and optical isolation techniques for individual molecules or nanoparticles on the other, new processes developed by physicists known as "series" type processes (since they work atom after atom) have, at the moment, expanded the methods used in the laboratory. These techniques enable scientists to study and manipulate an isolated nanometric object and at the same time to produce the smallest objects that man can create: the organized and stable assembly of some atoms and molecules. Beyond its technical use, this is about exploring a new domain where quantum effects call into question the established principles of solid state physics which govern all of our current industrial processes.

The isolation methods used for a unique nano-object create new opportunities for probing, measuring and manipulating the subject on a molecular scale. When an individual nanometric object is being studied, the information obtained is, by definition, free from all effects of mass. It has become possible not only to measure the behavior of an isolated nano-object and to control its behavior in certain cases, but for the first time it has become possible to directly observe the dispersion of local molecular parameters between two nanometric objects and to follow the temporal fluctuations of these parameters without having to synchronize the fluctuations of one of the objects. This approach is renewing our perception of disorder in the condensed object. It simplifies comparisons with theoretical models used in physics and allows for a better understanding of the interaction of objects with their environment.

Nanotechnology and nanoscience are interested in objects with the dimensions of an atom, in other words, the size of 1 nanometer (nm), i.e. 1 billionth of a meter or 1 millionth of a millimeter (10^{-9} m). More precisely, a nanometer is four times the diameter of an atom. In comparison, the size of a virus is 100 nm, and the size of a bacterium is a micron or 1 thousandth of a millimeter which corresponds to 1,000 nm. In addition, a living cell is close to 50,000 nm in size and a grain of sand is approximately one million nanometers.

In a grain of sand measuring 1 mm long there would be approximately 10^{18} atoms. Proportionally, this would be similar to a cube with a side measuring one kilometer filled with 10^{18} marbles with a diameter of 0.25 nm. To try and position the atoms one by one with the help of a device derived from a STM or AFM at a rate of 1/1,000 second per atom would take 30 million years to complete.

The process of external assembly atom by atom is therefore not compatible with the manufacturing of industrial objects.

In 1986 Eric Drexler, in his book *Engines of Creation*, laid the foundations for a theory that was largely controversial at the time, i.e. the principle of the assembly of self-replicating entities.

3.2. From the manipulation of atoms to nanomachines – the concept of self-assembly

The strength of molecular nanotechnologies (MNT) resides in two key ideas:

– the first idea is to implement ways of building materials by the positional assembly of atoms or molecules. This first principle can lead to the creation of new substances not necessarily available in nature and which cannot be produced using existing processes such as chemistry. The information technology used in molecular modeling is an essential tool for the design of these artificial molecules and materials. It enables scientists to verify the potential existence of these artificial molecules and materials and to validate their stability. In fact, all chemically stable structures and those which can be used to make models can be manufactured. This means that all the molecules can be broken down, their components can be reassembled into another molecule and then into macroscopic objects, which make up the first machines capable of artificially building on a molecular scale. The first nanomachines capable of building other nanostructured products on an atomic scale were thus created. These first nanorobots are called "assemblers";

– the second idea is that nanorobots are capable of making copies of themselves; copies which will in turn copy themselves and so on. The exponential growth of these tiny machines will allow scientists to create a product on a human scale from a macroscopic or nanostructured product in a few hours and with no waste, i.e. an unparalleled performance in comparison to processes which are used today. In theory, it will be possible to create complex structures of large dimensions with an unequalled precision: atomic precision is as robust as diamond or other diamondoids. This is what is known as molecular manufacturing. It will undoubtedly form part of the greatest opportunity regarding durable development. The principle, even if it seems rather futuristic, does have an undeniable support: the history of life sciences. Nature has been doing this for more than 3.5 billion years; the feasibility has therefore been established.

This is the essence behind Eric Drexler's theory and his *Engines of Creation* for which the Foresight Institute and the Institute for Molecular Manufacturing have become leading ambassadors.

In order to manufacture a useful macroscopic object it is necessary for these molecular machines to be able not only to duplicate themselves, but also to manufacture things other than themselves, i.e. other structures. The first version of an assembler would be a machine containing some millions of atoms and would integrate a manipulating arm allowing for the positioning of atoms one by one to any desired area. With some copies of this basic assembler, it would be possible to construct other more evolved nanomachines and it would also be possible to create the first self-assembled artificial objects on a molecular level. MNT research programs are currently exploring two ways of elaborating this first assembler: by direct construction with the help of STM and AFM microscopes mentioned earlier, but limited to simpler structures (up to a size of one million atoms at most), and by self-assembly through genetic engineering which exploits modern tools by the manipulation of pieces of DNA (see section 4.3.2). Genetic engineering through the use of restriction enzymes, ligases and transfer vectors (i.e. "cut", "copy", "paste" and "displacement" tools) has allowed for the creation of a real molecular programming language. Several laboratories have succeeded in manufacturing nanostructures by creating and joining several DNA strands, opening the door to the study of a first assembly made up of pieces of DNA.

To create this type of machine is one thing, to produce it industrially for practical use is another. Even if the problem of the feasibility of these objects could be resolved, we still need to make sure that these nanomachines function and resolve any engineering problems as soon as the first assemblers become available. This will, in any case, be a long process.

"By manipulating the molecules one by one, as we do today with scanning probe microscopes, it would take a prohibitive amount of time to create anything visible. It is an unworkable business. However, what we can do with this type of instrument is to construct small complex objects made up of some thousands of pieces which themselves would be molecular machines. If we had one molecular machine capable of producing another, productivity could become reasonable. The product is macroscopic as is the machine. If you use one machine to produce another, you will get two, which in turn become four and so on. This process speeds up very quickly until it becomes possible to work with some kilograms or some tons which would enable you to go on to the industrial level." Eric Drexler christened these engines of creation "assemblers". An assembler is a nanosystem connected to two mono-atomic probes. It is a machine which is capable of producing anything, including itself. The resemblance to living organisms is striking.

An assembler would need some thousands of pieces the size of a billionth of a meter, in other words, a nanometer and would be as complex as a living cell. To build one of these instruments would take an almost divine knowledge comparable to that which was needed to create life from chaos.

Life has effectively designed engines the size of a millionth of a centimeter. The cilia or flagella of bacteria are small helices, even if they are more the shape of a corkscrew. The cilia or flagella are the locomotor organs of living cells. These cilia and flagella look like a hair which is made up of an extension of the cytoplasmic membrane surrounding a formed structure of nine pairs of microtubules and forming a circle around a central pair of these microtubules (the "9+2" structure). Cilia and flagella which are found in almost all animals have the same structure.

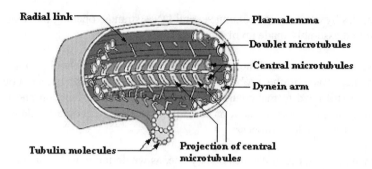

Figure 3.3. *A nano-engine: the flagella of a bacterium*

At the root of each cilium there is a tiny engine whose diameter is 350 times smaller than that of a human hair. This engine functions thanks to the difference in electric potential between the interior of the bacterium and its environment. It is the engine which activates the rotation of the flagella and which makes the bacterium move. It is therefore imaginable and, indeed, possible to apply this same idea to technology in the same way as nature did before.

From a scientific and technological perspective it is always risky, sometimes even dangerous to want to make too many predictions, particularly regarding the subject of living systems. Biological evolution is a slow and progressive process which advances in small phases. Remember that the history of living systems is a long one which has been going on for approximately four billion years and that this evolution is not yet complete. The creation of artificial objects by man cannot work against a progressive process even if the observation and comprehension of living systems allows for a significant acceleration in the development of artificial systems, i.e. the notion of time constraints where it is not possible to create something in a few centuries or decades which nature has been doing, indeed has been doing better, over several millions of years. The history of science seems clear whenever discoveries are analyzed in retrospect. However, 10 or 20 years later it is still difficult to admit at what point simple principles have been able to solve an unsolvable problem when it is actually about trying to understand these principles in the future on a medium term. The major point of contention amongst numerous scientists and Eric Drexler undoubtedly comes from how, rather than what. It is this reassurance, this unbreakable confidence that Drexler shows in his vision and in his mastering of an implementation plan removing all uncertainties, which disconcerts, indeed irritates, the scientific community, whose precision has always been based on extreme caution. However, it is extremely likely that the self-assembly of molecular systems in larger objects will become the cornerstone of the technology of the future.

According to the most passionate and optimistic defenders of molecular nanotechnology, scientists do not have the scientific knowledge required in order to produce and build a molecular assembler. Drexler himself admits that designing a viable assembler is one of the biggest engineering challenges in history: "It is about systems made up of several hundred thousand atoms, maybe more, with numerous mobile parts and numerous sub-systems which would carry out different tasks. Certain mechanical parts would push or pull something, whilst other parts would have to be able to grab then let go of something. Other components would be capable of creating a chemical reaction at a precise moment at a given place. There would also be elements which would supply raw materials and break them up in order to have the correct piece at the right place or at the right moment by blocking contaminating agents from the surrounding environment. The difficulty does not raise the question of scientific or technical feasibility, but rather the development of the production processes used. We should not wait until one person coordinates everything. Going from a plan to its implementation is an immense project, such as walking on the moon for example".

The principle of feasibility regarding the self-replication of assemblers is today the center of the debate concerning the industrial reality of molecular nanomachines which this latest technology has to offer. The main argument lies on the premise that all living objects, including humans, have been created in such a way, i.e. as molecular nanomachines. Let us take the example of ribosomes which barely measure some thousands of nanometers cubed. These mini factories produce all of the Earth's living proteins by combining pieces of RNA. Infinitesimal, ribosomes manufacture objects much larger than themselves by following a "plan" or a "program", that of the genetic code. However, the critics of this plan have responded to this by stating that there is a significant difference between the production, organization and control of a group of programmable assemblers and what happens in nature: the molecular system of living beings is extremely specific in its functions. It has been organized into specific paths for millions, indeed, billions of years and has been operating in symbiosis with a very large number of other sub-systems in an extremely complex framework which has been cleverly designed and of which, up until now, little has been understood (see section 3.4).

3.3. From the feasibility of molecular assemblers to the creation of self-replicating entities

"Heads without necks appeared, the naked arms did not have shoulders and solitary eyes wandered without a face. Feet were seen trailing countless hands." Then one day, by chance, the elements necessary for the construction of a being came together and with time the most poorly arranged combinations disappeared and only harmoniously assembled creatures survived. In the 5th century BC, according to Empedocles of Agrigentum[1], the first living being was born.

Empedocles' primitive theory considered that man and animals were developed from a previous form. This remains at the center of our conception of life on earth: certain molecules self-arrange themselves in water and form chemical structures, i.e. automated species capable of creating other molecules to generate structures of themselves. These chemical automatons can therefore produce more of themselves by themselves. This is what we call self-replication. As this replication process continues slight assembly risks appear, giving rise to automatons which are more apt at reproducing themselves. These automatons therefore make up the dominant species. This is evolution. Self-replication and evolution are the two properties which characterize the passage from inert matter to the world of the living.

It is never easy to clearly and simply explain the contents of a scientific dispute, particularly when it concerns the areas of physics and chemistry where the language, models and examples quickly become complex and tedious for non-specialists in these fields. It is nevertheless important to reflect on a moment of the dispute, largely covered in the media and in the scientific community in the USA, which saw Eric Drexler disagreeing with Robert Smalley concerning the feasibility of molecular assemblers (the origin of MNT). Today, the scientific community is still divided on the industrial potential of this technology.

Molecular assemblers would enable scientists to obtain biological self-replicating entities. It would also lead the way to decisive progress in the production of advanced technological objects. It is the ultimate stage in the advent of nanotechnologies which would largely overtake (as much in practical opportunities as in potential dangers) the nanotechnological applications devoted to the manufacturing of traditional industrial products and materials, such as carbon nanotubes which will be introduced later in this chapter. The traditional applications are the subject of ambitious research and development plans all over the world and their industrial potential has already been implemented into the production lines of large companies. These traditional applications are part of domains accessible to today's technology. They also belong to the foreseeable future. This is still not the case for self-replicating molecular assemblers. Eric Drexler has found the feasibility of his theory being contested by one of modern chemistry's leading figures.

---

1 Greek philosopher and scholar who believed that everything is made up of four main elements: earth, air, fire and water. Empedocles said that a change leading to the creation of new matter is impossible, the only possible changes are those which are produced through combinations of the four existing elements.

Richard Smalley, a distinguished nanotechnologist who won the Nobel Prize for Chemistry in 1996 for the co-discovery of fullerenes, has always issued extreme caution regarding the "myth" of MNT.

This polemic is far from the simple scientific-philosophical debate. It directly affects the destination of significant public funding in support of initiatives and other plans concerning nanotechnology in the USA and throughout the world. Neither industrialists nor taxpayers want to see these funds being distributed to research whose end result remains too distant in the future.

Scientists who grouped together in support of MNT criticized Richard Smalley for having used a somewhat fallacious or exaggerated argument to demonstrate the practical impossibility of implementing self-replicating entities which make up the essence of nanotechnology. These researchers in fact believe that the advanced conjecture on the limitations regarding the chemistry of enzymes would introduce some disproportions.

After having shared his iconoclastic and popular vision of a "programmable chemistry" in *Engines of Creation*, Eric Drexler developed a detailed theory in 1992 in a second, more technical, publication entitled *Nanosystems*. In it he puts forward the basis for the implementation of a production system on a nanoscopic scale or a programmable synthesis of diamondoids. This vision was christened Limited Molecular NanoTechnology (LMNT) in order to distinguish it from the generalization of the concept to a synthesis of all chemical substances which can be modeled.

> Largely inspired from molecular biology, studies in the domain of original nanotechnology have used the bottom-up approach. This approach relies on molecular machines assembling blocks on a molecular scale in order to produce sub-products and then finished products which include the manufacturing of these molecular machines. Nature has shown that such machines can easily be produced in large quantities for a small budget.
>
> It is possible to imagine placing each atom in a precise location in which it would either take up an active and defined role, or a structural role, i.e. become a material of construction.
>
> Such a system (in materials physics the term phases is used) would not be comparable to a gas or to a liquid in which molecules can move randomly, and would neither be comparable to a solid in which the molecules are fixed. In fact, it is about a new type of phase which the pioneers of MNT have named machine phase which is capable of authorizing molecular movements, i.e. degrees of freedom following the examples of liquids and gases, or showing qualities of given mechanical resistances which is one of the properties of solids. Molecules of such a phase exclusively constitute active mechanisms on a molecular scale.

The fear of nanorobots escaping from the control of their creators and attacking the biosphere, i.e. the risk of global ecophagy, rapidly sparked the imagination of science fiction authors and, more generally, the media. All sorts of applications and catastrophic scenarios have been touched upon. The majority of them have contributed to the marginalization of new scientific theories, even the most rigorous in this field.

It was towards the end of the 1990s when the American government was inspired by the incredible potential of this theory and when the first national initiatives, which were largely subsidized, came into being. Other states such as Japan and Taiwan, for example, quickly followed suit.

Many scientific disciplines then discovered that they used nanotechnology in different forms and were candidates for these programs. It was from that moment that the gap between "traditional" nanotechnology (based on industrial materials such as nanotubes and the continuing development of top-down miniaturization processes especially in the silicon industry) and molecular nanotechnology, which introduced the idea of self-replicating assemblies, started to widen. Traditional nanotechnology was able to justify predictable and rapid industrial repercussions and largely interested industrialists. It was a respectable area of research and development. A new technology concerning production with numerous programs, such as the pioneering study project at the Xerox Research Center in Palo Alto, had just been created.

Molecular nanotechnology, however, was, and for the moment still is, too revolutionary and still too far removed from the big leaders of research and development. It has been virulently criticized by scientists who were working in the area of nanotechnology, as well as by directors who were in charge of managing the funds. These investors were afraid of a sterile waste of resources in what they qualified as a doomsday scenario or an end of the world initiative. From that moment, numerous rumors, which came from the US media and scientific community, criticized a lack of clarity in the policy for supporting nanotechnology programs. With a massive investment increase at the beginning of the 1990s, and especially the share allocated to "long term research for the needs of Defense", certain analysts believed that the most ambitious research in the field of molecular nanotechnology had already got underway.

In September 2001, Richard Smalley published an article in *Scientific American* entitled "Of chemistry, love and nanorobots" with a sub-heading "How soon will we see the nanometer-scaled robots envisaged by Eric Drexler and other molecular nanotechnologists? The answer is quite simply, never". In his article, Smalley proclaims that chemistry is not as easy as Drexler makes it out to be, that atoms cannot simply be "pushed" one against the other in order to get the desired molecules and chemical reactions, but that their environment has to be controlled in minute detail. In order to back up his argument, Richard Smalley started by illustrating Drexler's concept by devising a system of "magic fingers" evolving in a working nanozone and manipulating the atoms one by one. Then, from this premise he concludes that such "fingers" would be too big to evolve in such a small volume and would be too "sticky" to release the atoms to the desired location.

> Catalysis, or the miracle of enzyme chemistry, which is one of the main principles of physics, wants all systems to adopt a structure in which internal energy is as small as possible. In the normal or stable state of a molecule the atomic nuclei and electrons are so close to one another that the underlying laws of quantum physics allow them to be there. In such a state the molecule can therefore be considered as an uncharged electric battery. A traditional estimation of quantum chemistry considers the atomic nuclei as fixed with mobile electrons flowing around this nucleus. Scientists also consider the electronic energy of the molecule. The total energy contained in a molecule is less than the total energy of each atom that makes up the molecule if we consider the atoms as being independent. This, in turn, explains the stability of the molecule. In order to cut the molecule into pieces energy must be supplied. On the other hand, the assembly of a molecule from isolated atoms should release some of the atoms, this releasing of atoms is a spontaneous process. In order to force atoms or molecules to interact they need a supply of activation energy. The new molecule restores all, or part, of this energy. To resolve this problem living cells have adopted a rather clever process, that of catalytic action (i.e. dissolution). According to a configuration which shows the group of atoms where the bond is likely to be established, it is rare that two molecules capable of joining together randomly meet from their trajectories. The collision is therefore in vain. The cytoplasm of a cell is full of specialized molecules (proteins; see section 3.4.1) whose conformation creates "niches" at the bottom of which other smaller molecules can be found on the condition that they possess the complementary conformation which corresponds exactly to this niche. For each chemical reaction required for metabolism, the cell produces a mass of proteins called enzymes whose conformation allows several given molecules to move themselves into a favorable position so that their union into a new assembly can take place. This catalytic action enables atoms and molecules to free themselves from the activation energy.

In April 2003, Eric Drexler addressed an open letter to Smalley saying that Smalley's illustration of the concept of "magic fingers" was an open attack on Drexler's work. He claimed that he never proposed anything of the sort and accused his colleague of having deliberately introduced confusion and worry among public opinion. Drexler did not get a response. He then wrote a second letter in July 2003 reminding Smalley that he himself had twice in 1999 and 2003 put forward the possibility of constructing objects atom by atom and that from this fact he considered the discussion as closed. These letters gave rise to the famous four part debate, which has since been the subject of many publications.

In response to Drexler's two letters Smalley recognized that his illustration of "magic fingers" was part of a somewhat slightly abusive antithesis. He also accepted the premise that something living, such as an enzyme or a ribosome, (components of living cells; see section 3.4.1) would be capable of creating from molecular chemistry the exact same thing on a nanoscopic scale, but he insisted that it would only work under a certain condition: in water. Suggesting an unusual alternative, that the nanofactory of his colleague contained the components of a complete biological system, he developed a new counter-argument in describing the limitations of chemistry in aqueous surroundings and concluded by asking the question "Do you really think that it is possible to implement a variant to do enzyme-like chemistry of arbitrary complexity with only dry surfaces and a vacuum?"

Drexler would retort in his book *Nanosystems* that the chemistry of dry surfaces and a vacuum, which he named machine-phase chemistry, can be very flexible and rational given that by maintaining a molecule in a spot generates a strong catalytic effect. He points out that chemical vapor deposition systems are examples of "dry chemistry" and highlights that positional control naturally avoids the majority of secondary reactions by obstructing unplanned conjunctions between potential reagents. In other words, the production of undesirable substances can be avoided.

Smalley answered Drexler's latest letter with a direct attack on the chemistry of dry surfaces and a vacuum. In his presentation he stated that chemical reactions must be absolutely controlled in a multidimensional hyper-space and due to this fact cannot be carried out by elementary robotics. Once again he developed his convictions regarding the chemistry of enzymes: "even if enzymes can operate a precise and reliable chemistry, they can only do this in an aqueous environment".

A technical analysis of the debate might have since shown an area of doubt in this last assumption likely to restore the basis of Drexler's theory (or in any case not invalidating it) and therefore the theory of the feasibility of MNT, i.e. molecular assemblers and self-replicating entities.

Regardless, in this field of science the utmost scientific caution is essential. Nothing has proven that one day we will be able to master the transition between the organic and the living. We are just starting to understand how the cell works. Our knowledge regarding the formation of tissues or organs (i.e. how can hundreds of billions of specific lines following a perfect order in time and space come from one cell?) still remains very incomplete. We do not focus enough on the system logic which governs the execution of complex programs such as the development of an evolved living being, for example. Nothing has told us that one day we will be able to describe all this information in terms of physics and chemistry.

3.4. Imitating nature – molecular biology and genetic engineering

From the most miniscule virus or bacterium to the most evolved mammals, every biological entity possesses a genome which contains the exhaustive description of the species. Nature has defined and implemented its own complex description of living organisms of molecular detail on a macroscopic scale.

This powerful and natural aid mentioned in the previous paragraph enables the cell mechanism to assemble with atomic precision all the proteins which will equip biological species with their properties inherited from evolution. The impressive nanotechnological model of nature is a mass production system organized on a molecular scale with an almost infinite diversity.

This mechanism has already been in use for several billion years and will carry on functioning well beyond our existence. It forms part of one of the most fabulous scientific hopes and revolutionary technological opportunities. Nanotechnology is a fascinating challenge for mankind. It is about building and manipulating structures which are too small to be seen on a human level. Nature has demonstrated that the potential impact of these "invisible machines" is extraordinary.

This book, which is mainly devoted to information technology and its microelectronic components, with a chapter devoted to molecular biology and genetic engineering, might appear here as irrelevant digression. However, artificial barriers between scientific disciplines have become less pronounced in the area of the infinitely small. These barriers were established throughout centuries because simplifying (and also reducing) models allowed for the understanding of concepts on a macroscopic scale by offering sufficient ways to understand and work at this level. However, specializations did not exist on the atomic level.

From the bacterium to the carbon nanotube transistor and diamond, the same basic building block of carbon is used. Only the application, or even the utilization of this building block, will distinguish it on a macroscopic scale. Nanoscience and nanotechnology form part of unifying disciplines. Progress or discoveries in one of the application areas will have major consequences for the other application areas. As such, the view of the solid-state physicist on bacterium and living cells certainly opens interesting applications to new models of computers.

"Every cell is born from a cell" is what Rudolf Virchow stated in 1855. The principle that no spontaneous generation exists marked the beginnings of modern biology.

Cells do not develop from inanimate matter and cannot be created out of nothing in the laboratory. Each cell comes from the division of a mother cell and this process is still part of one of the most phenomenal mysteries of life. How can a cell (whether it is a simple bacterium or a much evolved human cell) produce another cell without losing any of its characteristics in the process? How can a succession of divisions from only one fertilized egg result in the formation of such a complex creation as an animal or man, made up of billions of functioning cells following a perfect organization? Molecular biology and genetic engineering have found some answers to these questions. This knowledge will undoubtedly benefit the development of new types of information systems when silicon technology finally reaches its limits, whether it be a question of DNA computers or molecular memories using organic or biological molecules to code, store and retrieve information, all of which will be dealt with later in this book. Maybe one day genetic engineering will contribute to the success of molecular technology and the future of our computers.

3.4.1. *When nature builds its own nanomachines*

Gregor Johann Mendel (1822-1884) defined the gene as a cell factor which determines a given property (of vegetables at that time). He had already understood that all the cells of an organism contained genes. The set of genes of a cell is called a genome. In the majority of organisms this genome is made up of several chromosomes located in the cell nucleus. For example, the genome of a human cell is made up of 23 pairs of chromosomes, each pair being composed of a paternal chromosome and a maternal chromosome. It is estimated that humans have no less than between 50,000 and 100,000 different genes. In bacterial cells, which do not have a cell nucleus, the genome is often composed of only one chromosome which has between 3,000 and 5,000 genes. Certain bacteria contain additional smaller chromosomes called plasmids which contain only a few dozen genes.

> Viruses are not cells but the hosts of cells. They are uniquely made up of one viral chromosome and an envelope, and contain anything between several genes to several hundred depending on the virus. In order to multiply themselves, the virus introduces its genes into a given host cell. The cell is therefore "reprogrammed" and starts to produce the virus. Although viruses only colonize one specific organism in general, it is believed that viruses play an important role in evolution because they can pass genes from one species to a more or less similar species.

Genetic information is stored in each cell in the form of DNA (Deoxyribonucleic Acid). Each molecule is a giant molecule of filament-shaped DNA. The chemical components of DNA are four nucleotides called adenine, cytosine, guanine and thymine which are noted as (A), (C), (G) and (T) respectively. The sequence of nucleotides in a DNA molecule determines its genetic information. Genes are in fact fragments of DNA whose variable length is the size of between one thousand and some tens of thousands of nucleotides.

In 1953 two researchers, Watson and Crick, discovered the famous double helix structure. DNA is the wound shape of a spiral staircase whose two banisters represent the two chains or strands of DNA. Genetic information is stored in double (i.e. on each strand). Nature has invented a magnificent mechanism of error correction (close to that used in the software industry) avoiding the loss of information during gene manipulation. In fact the nucleotides, which are nitrogen-based complementary pairs, pair up (A with T and G with C) and due to this are always located opposite one another, each strand therefore expresses the same code in two complementary forms.

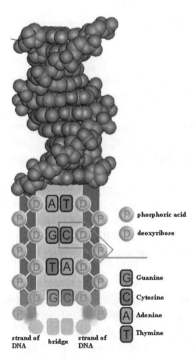

Figure 3.4. *In fact, DNA is a useless molecule as such. It is not mechanically interesting (unlike Kevlar for example), neither is it active like an enzyme, nor colored like a dye. Yet it possesses something which industry is ready to invest a large amount of money in: it is capable of running molecular machines which make up ribosomes*

DNA ensures the transmission of hereditary information to cells during their multiplication and, consequently, the transmission to descendants of the strongest living beings. During cell division genetic inheritance is copied using enzymes (i.e. protein playing the role of the biocatalyst).

If DNA is the "software" or construction plans of the cell, proteins which are the biological components are the materials that enable the physical creation of the cell. Each cell contains thousands of different proteins which take up and perform specific functions. They mainly ensure the functions of:

– messenger or signal (these are the hormones);

– receptor (which detects messages or signals);

– biocatalyst (enzymes taking part in the transformation of substances);

– construction material (to give shape and structural support to the cell);

– canal or pore (in the cell membranes);

– antibodies (the immune system's defense substance);

– toxin (protein-type poison).

The role of proteins is therefore chemical (the transformation and transfer of information) or purely mechanical. Certain proteins push and pull, others are fibers or supports, and the proteins, which are part of some molecules, produce excellent movements. Muscle, for example, is made of clusters of proteins that attach to fibers (also made up of proteins), pull the fiber, and then re-attach further along the fiber. These machines operate with each movement our body makes.

Despite their different functions, proteins belong to a unique category of substances because they are all formed from the same constituent elements: the 20 aminoacids. These aminoacids are dispersed one behind the other just as in a pearl necklace. According to the sequence of theses acids the necklace coils up into a predetermined three-dimensional structure, therefore giving a specific protein an assigned function.

The genes describe the nomenclatures of the aminoacids used and the production range, i.e. the order in which each aminoacid must be assembled in order to form a given protein. The gene thus determines the sequence of the aminoacids which will make up the protein chain. The genetic code enables us to predict the formed protein from a particular gene. The production of a given aminoacid always depends on the combination of three components of the DNA.

Nature has curiously standardized its genetic code. Whether it concerns animals, plants, unicellular organisms or even viruses, every organism uses the same genetic code. This is why it is possible to transfer genes of fragments of DNA to different species of organisms, enabling them to manufacture proteins which are foreign to the species by using new manufacturing programs.

This manufacturing occurs within the cell in the real protein nanofactories called ribosomes. In fact, to produce proteins these ribosomes only receive working copies or duplicates of the genes. Contrary to the gene itself, these copies do not contain any DNA but a closely related chemical substance, ribonucleic acid or RNA. This acid has only one strand, unlike DNA which has two. Transcription is the name of the process used for copying DNA into an RNA messenger. To produce a protein, a ribosome reads an RNA messenger molecule (as a microprocessor would read a register or a word from memory). It always reads three components of RNA at the same time. The four components of RNA are noted as (A), (U), (G), and (T). The three components that are read determine which of the 20 aminoacids will be introduced in this region in the protein molecule. An analogy with the software industry would be that the protein chain is an assembly of three-letter words made with any combination of the four-letter nucleotide alphabet, i.e. (A), (U), (G), and (T). Enzymes (the catalysts) then arrange the appropriate aminoacid in the protein chain which is being formed. The ribosome then processes the three components of RNA and adds the appropriate aminoacid. This process is repeated until the protein is completely formed. The resulting protein chain coils up into a bell shape whose three-dimensional structure depends on the sequence of aminoacids. This structure determines the function of the protein. Nature translates purely genetic information into a biological function (the process of protein synthesis has adopted the term translation).

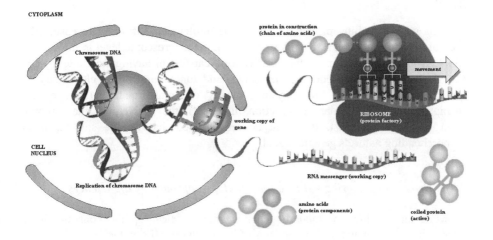

Figure 3.5. *Ribosome: the protein factory*

It is not possible to conclude this description of the industrial chain which takes place within living organisms without mentioning the remarkable mechanism of gene switching. At each instant, a large number of simultaneous chemical phenomena develop in the cell: the transformation of molecules, transfer of energy, transport of substances, cell construction and assembly and the repair of genetic inheritance. Before and during cell division an important and complex transformation also occurs: chromosomes have to be split in two, carefully open out and transfer in identical form to two daughter cells so that they have a chance of survival. However, this is not enough in the case of a complex organism. In this case, the cell adopts a "social" role within the organism, i.e. it must undertake one or more specific functions such as the production of a hormone or the production of mechanical work (a muscle). These macroscopic mechanisms must be coordinated in the same way as the microscopic functions which guarantee the existence and even the survival of the primary cell. Nature has implemented an absolute control system of all processes: the cell has an arsenal of regulation mechanisms and in particular the ease with which it can selectively activate or deactivate genes. This is what is known as the control of gene expression. In a complex organism cells are generally devoted to one function. In this type of specialization numerous genes, whose cell is not used for carrying out any functions or which are not needed for survival, are deactivated. However, genes will produce the most frequently required elements for the cell by obeying transcription factors: if the cell needs a protein, these transcription factors will settle in specific regions of the DNA (known as control regions) and will thus activate the appropriate gene. This activation leads to the process of transcription, i.e. the production of a working copy of the gene used by the ribosomes which will authorize the physical production of the protein in question.

This gene switching principle needs to be implemented in order to ensure the control of the production of complex objects on the macroscopic level in the case of MNT. If the assemblers mentioned earlier were capable of having local intelligence (i.e. the capacity of reproducing themselves and functioning on an elementary or unitary scale) and had the physical means of transformation (storage and manipulation of molecular construction materials), the problem of macroscopic control would still exist. It should be remembered that we are still far from comprehending nature's genetic model.

3.4.2. *Genetic engineering – the nanotechnology approach by life sciences*

Molecular biology or modern genetics aims to clear up the constitution of genetic information and the mechanisms of its cellular use. From the 1950s onwards, experiments carried out on organisms with a relatively simple structure (bacteria and bacteriophage viruses) promised interesting opportunities for molecular biology.

However, approximately 15 years later it was the area of genetic engineering which allowed for a clearer understanding of complex cell systems and paved the way for recombinant genetic manipulations.

Genetic engineering came about in 1973 when two American researchers, Stanley Cohen and Herbert Boyer, succeeded in implanting an assembled DNA molecule inside a bacterium. The constructed molecule was then passed on to other bacteria descending from the initial bacterium, i.e. it became hereditary. Since then there has been continued development in experimental techniques which have already created interesting industrial and commercial applications.

Like a traditional repetitive manufacturing model, the tool box of molecular biology contains three types of instruments to ensure the following functions: "copy", "cut" and "paste". These functions can be ensured by enzymes coming from viruses, bacteria and other superior cells which will become the working tools of this industry. Amongst these enzyme-like tools it can be stated that restriction enzymes are the equivalent of scissors, ligases form the DNA glue and the DNA polymerase works like the copies of DNA.

Approximately 30 years ago, Werner Arber, a researcher from Basel, Switzerland, discovered that bacterial cells protect themselves from the penetration of foreign viral DNA thanks to restriction enzymes. This mechanism works in the following way: some DNA sequences (for example, GAATTC) are almost never dealt with or are uniquely present in a chemically protected form – in the set of genetic inheritance of Escherichia Coli, the reference bacterium that is only found in intestinal flora. One of the restriction enzymes of this bacterium, called EcoRI, recognizes the GAATTC sequence. After the G of this sequence the restriction enzyme will cut all viral DNA which is trying to penetrate the bacterium. It is this mechanism of self-protection which is exploited in the restriction enzymes.

Of course, not all genes are defined by an AGGCCT sequence (the site of action of the EcoRI restriction enzyme is the complementary sequence GAATTC) and cannot be sectioned from the genome by using the EcoRI restriction enzyme. Molecular biologists have undertaken a long, important piece of work which aims at determining the site of action of several restriction enzymes in a fragment of DNA. Fragments of DNA can be combined amongst themselves only if these restriction tools are available.

It is in this manner that scientists have been able to isolate, in more than 200 bacterial species, a large number of restriction enzymes capable of recognizing different sequences. These restriction enzymes are part of an essential tool in molecular biology enabling scientists to carry out recombinant genetic manipulation with remarkable precision. For example, a fragment of DNA cut by EcoRI only corresponds to another fragment cut with the help of EcoRI. Other enzymes, ligases or DNA glue are used to securely assemble the extremities on which the slightest trace of suture will no longer linger. This process of assembly by fragments of DNA is known as in vitro recombination.

The final process to be included is that of the massive reproduction of this recombined fragment by exploiting mass production workshops which exist within cells and bacteria. When foreign DNA is introduced by in vitro recombination into a bacterial plasmid (i.e. small chromosomes made up of anything from several to several dozen genes) the recombining plasmid will act like a natural plasmid when it is reintroduced into the bacterium. This means that the bacterial cell will multiply the recombined plasmid through the process of cell division, meaning that the plasmid will be present in descendants of this bacterial cell. Vectors is the term used for this recombining plasmid, and the bacterium which is endowed with one of these plasmids is called a recombining micro-organism. The genomes of bacteriophages (i.e. viruses which infect bacteria) can also be used as vectors.

Despite the polyvalent character of proteins, a technology based on the exclusive use of proteins cannot be considered viable from the point of view of engineering sciences. In fact, protein machines stop working when they are dehydrated, freeze when they are cooled, and cook when they are heated. It would be unrealistic to reduce the approach of nanotechnologies to only protein engineering. However, using protein machines to construct nanostructures made from materials which are more solid than proteins will be one of nanotechnology's big challenges. This idea is similar to the idea of flying an engine heavier than air at the end of the 19th century, or walking on the moon at the beginning of the 1960s, both of which received pragmatic scientific judgment. Future engineering could perfectly envisage a protein machine similar to an enzyme which would add carbon atoms to a precise region, layer after layer. These self-assembled (i.e. produced from the bottom to the top) atoms would form part of a very fine fiber of a diamond (mentioned earlier) which is several times more resistant than steel with a density less than aluminum.

3.5. From coal to nanotubes – the nanomaterials of the Diamond Age

In 1991 the Japanese scientist Sumio Iijima, whilst working at her electronic microscope at the NEC Fundamental Research Laboratory in Tsukuba, Japan, observed the existence of foreign nanoscopic strands in a soot stain. Made up of pure, regular and perfectly symmetrical carbon, just like a crystal, these long and incredibly thin macromolecules would soon be known as nanotubes. They have been the subject of intensive research and study programs ever since.

More recently, nanotubes have become a material of predilection for future industrial applications.

If Sumio Iijima was the first person to observe a carbon nanotube, she certainly was not the first to work with it. Probably our ancestors, some 500,000 years ago, already produced tiny quantities of carbon nanotubes in the fires that they lit in order to cook their food and protect their caves. In fact, split by the heat effect, carbon molecules can see their atoms recombine as they can in soot. Certain atoms create tiny amorphous drops, others form geodesic structures, i.e. spherical shapes with many sides such as a soccer ball, and others are reformed as long cylinders (relatively long in comparison to their tiny diameter), and these long cylinders are the nanotubes.

Up until 1985 the only crystallized forms of pure carbon that were known were graphite and diamond. Diamond is the hardest, the densest and the least reactive of all the allotropes. Colorless and a semi-conductor, diamond owes all its properties to the arrangement of its atoms. Graphite is the most stable of the allotropes. Black in color and with a soft consistency, graphite is a good conductor. The conductivity of graphite layers is 10^{18} times more than that of diamond. Graphite exists in two forms: hexagonal and rhombohedric. The hexagonal layers are arranged in parallel and the forces existing between the layers are weak. The layers can slide at the same time, one on top of the other. This is where graphite gets its property of being an excellent lubricant from.

In 1985 three researchers, Robert Smalley, Robert F. Curl (from Rice University, Houston) and Harold W. Kroto (from the University of Sussex) discovered a new allotropic[2] form of carbon, the molecule C_{60}. This molecule is made up of 60 carbon atoms spread over the top of a regular polyhedron which is made up of hexagonal and pentagonal sides. Each carbon atom has a bond with three others. This form is known as Buckminsterfullerene or Buckyball and owes its name to the American architect and inventor Richard Buckminster who has created several geodesic cathedrals whose shape is similar to the C_{60} molecule.

More generally, fullerenes, which C_{60} is a part of, are a new family of carbon compounds. Non-equilateral, their surface is made up of a combination of hexagons and pentagons like the sides of a soccer ball. This arrangement gives them a closed structure in the form of a carbon cage.

---

2 Allotropy is the ability of certain bodies to present themselves in certain forms with different physical properties.

It was not until 1990 that Huffman and Kramer from the University of Heidelberg developed a production process enabling scientists to obtain these molecules in macroscopic quantities.

These processes have ushered in a new era for carbon chemistry as well as for the technology of new materials. Nanotubes were identified six years later in a sub-product in the synthesis of fullerenes. Fullerenes have mechanical properties equivalent to Young's modulus or an elasticity of 2×10^{12} Pascals which is 10 million times the elasticity of tempered steel. It also has a breaking strength 10 to 100 times larger than that of steel as well as a mass which is six times less than that of steel. Nanotubes also possess extraordinary properties of thermal conductivity and resistance to heat.

These nanostructures are sheets of graphite rolled into cylinders and closed at the two extremes. There are two types of nanotubes: single-walled and multi-walled. The extreme length of the nanotubes (the size of a millimeter) in relation to their diameter (of only a few nanometers) connects these nanostructures to hollow nanofibers, where the term nanotube comes from.

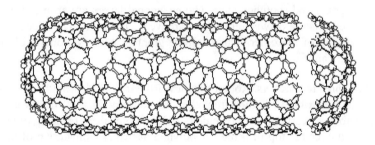

Figure 3.6. *A crystallized carbon nanotube, closed at its extremes (diameter is the size of a few nm)*

The first nanotubes which were observed in 1991 were multi-walled nanotubes (i.e. stacking similar to that of a Russian doll). It was only two years later when the first single-wall nanotube, with a wall thickness of only one layer of carbon atoms, was created. The properties of these structures were proving to be very impressive.

The secret of the astonishing stability of these tiny tubes is simply the strength of the bond between the carbon atoms. It is this strength which ensures the hardness of the diamond. The atomic arrangement of diamond is a tetrahedron with each of the four faces of the tetrahedron linked to a carbon atom. In a carbon nanotube the atoms are organized into a sheet of hexagonal honeycombed shapes, just like a thin sheet of a bee's nest which is coiled around itself.

Long before the discovery of nanotubes, scientists were able to industrially produce nanoparticles with interesting mechanical and thermal properties. These properties have been exploited in a wide field of applications. The industrial production of these nanoparticles was not carried out using advanced technology, but rather by a process of meticulous mechanical grinding of the substrate.

The manufacturing of carbon nanotubes today is achieved using three processes. In the three cases, it is a question of producing the soot which contains a significant amount of nanotubes. However, these three processes suffer from a common handicap: they produce, in bulk, a mix of different nanotubes of very different lengths with numerous flaws.

The first process was published in 1992 by Thomas Ebbesen and Publickel M. Ajayan from the NEC Fundamental Research Lab in Tsukuba. In this process, an electric arc of 100 amperes, which was created between two graphite electrodes, vaporized the extremities of the electrodes into a very hot plasmid. During recondensation, certain atoms recombined as nanotubes. The high temperatures that were used and the addition of metallic catalysts to the electrodes allowed for the creation of single-wall nanotubes (SWNT or SWNCT), or multi-wall nanotubes (MWNT or MWNCT) with some structural flaws.

The second method was the work of Morinubo Endo from the Shinshu University of Nagano. It is the principle of Chemical Vapor Deposition (CVD) which consists of placing a substrate in an oven at a temperature of 600°C and progressively adding a gas full of carbon, such as methane for example. By decomposing, the gas frees carbon atoms which recombine in the form of nanotubes. Jie Liu and his team from Duke University perfected the process by inventing a porous catalyst which converted almost all of the carbon from the supplying gas into nanotubes. In determining the catalyst particle models on the substrate, Hongjie Dai and his colleagues from Stanford University showed that it is possible to control the region where the tubes are formed. At the moment, they are trying to combine this controlled localization growth with the technology of silicon semi-conductors in order to create hybrid molecular electronic devices (see section 3.6). This process is the easiest to implement on a larger scale in the context of industrial production. It also allows for the production of long length nanotubes, which are required by fibers used as charges in composite materials. However, the nanotubes that are produced from this process are generally multi-wall with a lot of flaws and, as a consequence, are 10 times less resistant than those created by electric arc.

The third, more recent principle was put forward by Nobel Prize winner Richard Smalley and his team from Rice University. The method, which is based on laser ablation, consists of striking a graphite tube with powerful impulses from a laser beam in the presence of a catalyst. SWNTs are condensed onto a collector. This method is extremely efficient. Up to 70% of all the removed mass is transformed into SWNTs whose section can be controlled by varying the temperature of reaction. However, this method is by far the most expensive because it needs the use of power lasers which are very expensive to purchase and use.

We will experience a real revolution in the area of materials when carbon nanotubes can be produced at an affordable price and at the same time produce a charge that can be used in composites on an industrial scale. With the introduction onto the market of carbon fibers in composite materials 25 years ago, their performance was remarkable compared to the traditional materials of the time. Yet, their properties remain very ordinary in terms of elasticity, ductility and thermal stability in comparison to what carbon nanotubes have to offer.

The aeronautics and aerospace industry will more than likely be one of the main clients using nanotubes. A company which has been founded and dedicated to the production of nanotubes for the aerospace industry already exists. The production cost for nanotubes still remains incredibly high: between €20 and €1,200 per gram in 2002. However, while this cost will enable the use of these materials for the general public, the automotive, building, transport and sport sectors will be the principal customers of large volumes of nanotubes. A new industry, industrial nanotubes, will enter into an era of mass production. In this era of mass production, following a process of natural selection which has been repeated throughout industrial history, this market will be governed by profitability and economies of scale and will only have real openings for a small number of operators that exist amongst the current numerous producers of nanotubes. Nanotechnology will, in all likelihood, experience what economists call a shake-out, i.e. a major industrial restructuring accompanied by a significant social impact, possibly before the end of the decade. The Hitachi Research Institute estimated in 2001 that the market of nanostructured materials and associated processes should be worth $340 billion over the next 10 years. Although the current production of nanotubes still remains very low (from a few kilograms to a few dozen kilograms for the majority of operators in this market), today certain companies produce tons of carbon nanotubes per year. These more significant volumes of carbon nanotubes are exclusively for MWNTs which, unfortunately, do not offer the same legendary mechanical resistance as composites filled with SWNTs. The SWNTs are made up of longer fibers and, because of this, they have a more interesting mechanical potential. However, the current production method of these SWNTs is only capable of producing a maximum of a few hundred kilograms per year.

The extraordinary mechanical properties of nanotubes (mechanical elasticity, conductivity and thermal stability) have rapidly led to the most optimistic projections regarding the development of vehicles which are super resistant to shock, and buildings which are super resistant to earthquakes, as well as the development of microscopic robots and other futuristic applications.

However, the first industrial applications of these tiny tubes do not exploit any of these extraordinary properties. In fact, the first products took advantage of the electrical characteristics of carbon nanotubes. General Motors integrated them into certain components of car body-making using plastic materials. Car painting chains of the automobile industry have for a long time been exploiting the electrostatic effect on metallic parts. By applying a difference of potential between the component to be painted and the nozzle of the spray gun, this allows for a uniform and economic distribution of paint by efficiently concentrating the flow of paint coming from the spray gun. The inclusion of nanotubes in the plastic components to be painted allowed for this process to become widespread.

In the medium and long term, it is likely that applications with the highest added-value will rely on electronic properties on a nanometric scale. More precisely, it will be about exploiting the semi-conducting properties of nanotubes in molecular electronic devices. Carbon nanotubes could therefore adopt an interesting role by replacing silicon, but on a scale which is 10 to 50 times smaller, where silicon devices will have reached their limits and will not be able to be used. Experiments which have been carried out for more than a dozen years now show that it is possible to construct conducting wires and even functioning switching devices, logic gates i.e. the nanocomponents of a new generation of computer, by using carbon nanotubes incorporated into electronic circuits. These hybrid devices measuring less than 10 nanometers work much faster and consume a lot less energy than current silicon-based technologies (see section 3.6.1).

> The global potential of nanotechnology in the field of electronics has been estimated at a value of $300 billion per year for the next decade, to which an additional $300 billion can be added for the sale of integrated circuits (R. Doering, *Societal Implications of Scaling to Nanoelectronics*, 2001). However, if the macro market forecasts seem to be accessible in the field of nanostructured materials, they can become quite hazardous in a domain such as electronics, which rests close to breaking new concepts and radically new directions.

What is it about carbon nanotubes that make them so special for use in applications belonging to this new domain of molecular scale electronics? We have mentioned the geometry and the chemical composition of nanotubes which come from their mechanical properties. This curious arrangement also explains the complexity of the electron cloud of these atoms. Due to the size of nanotubes (we are entering the domain in which the laws of quantum physics reign), and also because graphite is a material with singular properties, nanotubes form a new class of conductors.

We have seen that electric conductors either belong to the class of metals or to that of semi-conductors, following the electronic population of the conduction band (see section 2.2.1). Graphite is one of the rare materials known as semi-metals, i.e. a category which is found in the transition zone between metals and semi-conductors. By combining the singular properties of semi-metals with the rules of quantum physics, in addition to the dimensions of these tiny tubes of matter, these objects will be excellent candidates for the implementation of properties which, up until now, have not been explored in applied electronics.

One of the laws of quantum physics, that of wave-particle duality, states that electrons act as particles and as waves at the same time. Electrons circulating around the wall of a nanotube can completely self-neutralize. Due to this fact, only the electrons with the correct wavelength continue to exist. Let us take into consideration the example of a sheet of flat graphite. Amongst all the electronic wavelengths or quantum states that can be observed on this sheet, only a small sub-set of them will continue to exist if this sheet is rolled into a thin twisted cylinder (i.e. a carbon nanotube). In other words, the nanotube acts as a filter.

Filtering a few of the quantum states of a metal or semi-conductor will change almost nothing in terms of the physical characteristics of the overall material. However, semi-metals are materials which are much more sensitive, which is where the interesting properties of our nanotubes come from.

In a sheet of graphite in which there is a high electron density (i.e. it is a material that conducts electricity), only the electrons close to the Fermi level contribute to conduction. None of the electrons of the carbon atoms in any other energetic state can move freely. One-third of the existing nanotubes simultaneously have the correct diameter and the correct twist structure so that the Fermi level can be included in their sub-set of authorized quantum states.

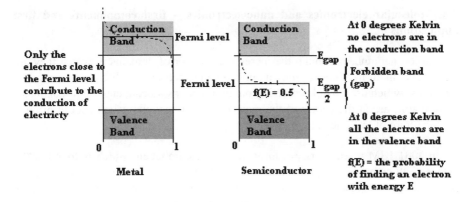

Figure 3.7. *Conductors, semi-conductors and semi-metals. Nanotubes form part of a new class of conductors*

The remaining two-thirds of nanotubes are therefore semi-conductors. This means that, as explained above (see section 2.2.1), their conduction band is empty at zero degrees Kelvin, meaning that they do not allow electrical current to pass without a moderate supply of energy which makes the electrons migrate from the valence band to the conduction band. This supply of energy can come from a flow of light or from the difference in electric potential which will move the electrons to the conduction band where they will develop freely.

The quantity of energy required depends on the width of the forbidden (or gap) band of the semi-conductor. As we have already seen, this forbidden band originates from electronic devices such as transistors. The diversity of the semi-conducting materials used, and therefore the parameters of their forbidden bands, allow for a wide range of components with different characteristics.

This controlled diversity is also possible in carbon nanotubes. Carbon nanotubes do not all have the same forbidden band since the section of the tube determines the authorized sub-set of quantum states, i.e. the energy levels of the valence and conduction bands. Thus, nanotubes with different diameters can act as metals (where the width of the forbidden band is zero) or can have similar characteristics to silicon with the existence of a forbidden band which is the size of one electron volt. The nanotubes can also authorize large numbers of intermediary forbidden bands whose widths are between zero and one electron volt. No other material currently known can control and adjust with such ease the behavior of its electron.

3.6. Molecular electronics and nanoelectronics – first components and first applications

Carbon nanotubes, through the incredible number of domains in which they are used, have, in all likelihood, brought applied nanoelectronics to life. These universal materials, which are capable of acting on command as a conductor, semi-conductor or insulator, are at the center of the majority of applications which should provide for the future of the silicon industry.

In 2001 IBM researchers put forward a basic component which is found at the center of all computers, a "NO" gate or a "reverser", created from a single nanotube. Since then, other researchers have been developing a collection of more complex devices which use these tiny tubes with extraordinary properties. In 2002 several laboratories were experimenting with nanotube transistors that function in the same way as conventional transistors, but whose performance is much better than that of their silicon counterparts. In the same year IBM announced that experiments from a nanotube transistor, which will be introduced later in this chapter, surpass the performances of the best silicon devices currently around.

Nevertheless, two major obstacles will have to be overcome by the nanotube electronics industry in order to ensure viable industrial opportunities. First of all, there is the problem of connectivity, because even if the manufacturing of a carbon nanotube transistor is possible in the laboratory, the production of a chip containing several million of these interconnected devices is far from being achieved. Once a technological solution has been reached in terms of the integration of a large number of these new devices on the same circuit, the challenge remains with regards to the development of an industrial process which will allow for the mass production of these devices at a minimum unitary cost. The strengths of the silicon industry have largely been dealt with. The basic technology relies on heavy investment (i.e. the elaboration of masks; see section 2.2.3), but also heavily relies on the cost of very large circuit series whose unitary cost per transistor is barely 2/1,000 of a cent (€). The electronic nanotube devices are made today in a craftsman style of manufacturing, component by component, similar to how the car used to be assembled in the days of Panhard and Levassor.

Without spreading doom and gloom, remember that scientists and engineers do not have the technology able to integrate the fundamentally different components of silicon technology, nor do they have the conjecture to design such a device (see section 5.7). To develop a universal computer similar to our current machines requires the mastering of three functional components: a switching device such as a transistor, a memory function, and this crucial method of connecting a very large number of switching devices and elements of memory. It will be the Achilles' heel of this new generation of machines as long as we consider a conceptual structure brought about by our current vision of information technology. However, considering the future as an extension of the principles and uses of current technology, which we believe today to be established and invariant, makes our prospective vision somewhat biased. This new generation of computers could rely on radically different structural concepts and could exploit new programming and operating principles in order to show the promising potential of the molecular components.

3.6.1. *Carbon Nanotube Field Effect Transistors (CNFET)*

Until now numerous research groups have worked on electronic switching devices made from carbon nanotubes, in particular the Carbon Nanotube Field Effect Transistor (CNFET). The CNFET uses a semi-conducting nanotube as a canal through which the electrons can travel in a controlled manner according to whether a difference in potential on the third electrode is applied or not.

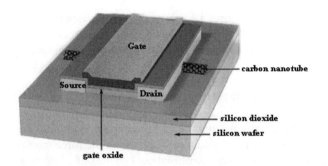

Figure 3.8. *Diagram of a cross-section of an IBM Carbon Nanotube Field Effect Transistor (CNFET) (source: IBN)*

Unlike the spintronic components mentioned previously, such a system has the enormous advantage of being able to operate at an ambient temperature and by following the functioning characteristics that are very similar to those of conventional silicon technology. The control current which is applied to the gate electrode allows for the variation in the conductivity of the source-drain canal by a factor of one million, indeed more since a conventional FET comes from the silicon industry. However, the dimensions of the canal's electrons enable the device to function efficiently and extremely frugally in terms of the quantity of electrons needed in order to function. It is estimated that a CNFET switching device uses nearly half a million times less electrons than its silicon FET counterpart. Electrical consumption is therefore reduced by such proportions. Theoretical studies consider that a real switch which would operate on a nanometric scale would be able to function at clock rates of one terahertz and more, i.e. more than a thousand times faster than our current processors.

In May 2002 IBM announced the creation of one of the best performing carbon nanotube transistors produced up until that point. IBM showed that carbon nanotubes, which are 50,000 times finer than a human hair, can outperform the unitary performances of the best silicon switching devices from that time.

IBM physicists and engineers experienced the strongest transconductrance (i.e. the opposite of the internal resistance of a component) obtained up until then with a nanotube transistor, or even with the best transistors coming from the silicon industry whose transconductrance turns out to be less than half, even by using lower electrical tensions. Beyond the exceptional results regarding transconductrance, this transistor demonstrated impressive properties such as much more straightforward responses in the switching of logic gates, from 0 to 1, or the reverse (the sub-threshold slope).

Such a CNFET transistor uses a single-wall carbon nanotube in a structure which is similar to a conventional MOSFET. The command electrode (gate) is placed at the top of the conduction canal and is separated from it by a thin dielectric layer (gate oxide).

The formula allowing the carbon nanotube to bond with metallic electrodes varies a lot depending on the research groups. This formula always relies on a hybrid process which combines traditional lithography, in order to create electrodes, with the use of very high resolution tools such as atomic force microscopes (AFMs) in order to localize and position the nanotubes. The idea of a largely automated mass production with billions of billions of transistors being produced at the same time is a long way away. This fact alone makes the silicon industry the leading force in the industry of current information systems. When witnessing the experimental device of Bardeen's, Brattain's and Shockley's transistor at the end of the 1940s, no one would have been able to imagine that the industry could one day end up with production processes on a scale that we are experiencing today.

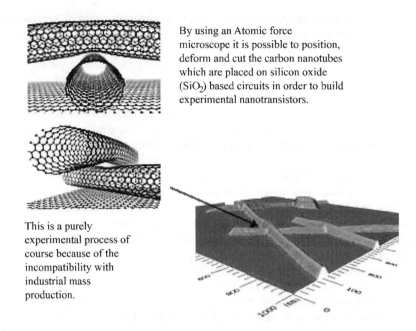

Figure 3.9. *An experimental process of positioning and deforming carbon nanotubes on a silicon dioxide (SiO₂) based circuit with the help of an atomic force microscope*

However, before thinking about more complex circuits which are based on nanotube technology, methods must be developed which enable the controlled growth of these tiny nanotubes in specific regions and according to specific positions, forms and dimensions. Researchers from Stanford University, and also other scientists from several other laboratories, have shown that by placing probes of nickel, steel or probes made from other catalysts onto a substrate, they can localize the conditioning of the growth of nanotubes. Another research group from Harvard University discovered a way to fuse carbon nanotubes with silicon nanowires enabling the interconnection of nanotube components with circuits from the conventional silicon industry. There have been many small steps taken towards large-scale use of these small carbon tubes such as active switching devices and conducting fibers for interconnections between components.

3.6.2. *Hybrid mono-molecular electronic circuits*

Today, the development of nanotube devices is part of the cornerstone of a new discipline: molecular electronics. Other approaches also exist. In each case it is a question of hybrid processes, i.e. the electronic circuit is created by interconnecting elementary molecular components (equivalent transistors), which are created from a single molecule by tiny metallic conducting fibers. Ultimately, the aim is to minimize microprocessors and memory.

Until now, none of these new methods have been able to resolve the issue of the coordinated assembly of a very large number of extremely small elementary devices and yet they claim to be part of an alternative to conventional semi-conductor technology, an alternative which would have to be industrially viable in order to continue the miniaturization process of electronic circuits (see section 2.3).

Several important researchers in molecular electronics have come up with a new plan for machines on a nanometric scale. Rather than interconnecting primary molecular components (interconnection being the major obstacle), they suggest integrating all the functions of a processor onto one molecular entity in order to reach the physical limits of the miniaturization of certain circuits for our future computers. The chemical limits concerning their production are, of course, the technological limits of their interconnection and integration.

In other words, the question "what is the minimum number of atoms needed for the creation of a given machine?" can lead to two types of solutions. The first solution relies on the conventional principles regarding electronic structure, which means designing circuits where each component would be replaced by a molecule. The second approach would consist of designing a circuit, i.e. the components and their interconnections, within one unique molecule. This is a rather radical method, which would free itself from the typology of electric and electronic circuits dating from the 1840s and which would explore the alternative to a functional mono-molecular circuit. Strictly speaking, such a circuit would no longer be restrained by the quantum behavior of the molecules and atoms as is the case with conventional circuits.

The results of experiments and calculations carried out at the end of the 1990s already showed that small electronic circuits could be created by using only one molecule. The use of a single molecule to create a machine (whether it is used to process information or to generate movement like a mono-molecular engine) demands a precise control of the quantum behavior of this molecule and also the ability to master its interaction with the environment. This futuristic engineering started the crossing over process of the areas of nanotechnology (where it is possible to integrate an electronic component into a specialized molecule) with quantum processing techniques (where it is possible to learn how to control the intrinsic quantum dynamics of a system in order to carry out a logic or arithmetic calculation).

3.6.3. *Organic molecular electronics*

This is an interesting alternative to carbon nanotube technology. Organic molecules have also shown that they possess the essential properties of a switch. The first mono-atomic transistors based on this technology have already been created. These forerunners, which are much smaller than silicon transistors, are part of the ultimate stage of miniaturization currently available (unless we go into the area of sub-atomic particles). They have largely taken their inspiration from nature. Furthermore, the pioneers of this discipline were extremely quick to gear their first studies to this area of science.

A new vision of molecular electronics grew from these observations. It is based on the use of a living material where the biological molecules are extracted from their natural context (for which they have been created) and which are re-used in a foreign and artificial environment with the aim of having a different use, i.e. completely "non-biological" uses.

For approximately 15 years researchers have been studying organic molecules and their potential applications in information technology, such as the use of biological molecules to encode, manipulate and find information, in other words to allow for the development of organic memory circuits. There are existing biological molecules whose two stable states of their atomic structure can be controlled; these states represent the logic states of 0 and 1 by benefiting from the photo cycle of these photosensitive proteins, i.e. their property of being able to experience a significant structural change when they are lit up. Such a protein changes state according to the light present and these logic states can be detected by an optic system.

Robert Birge from the W.M. Keck Institute Center for Molecular Electronics was one of the first pioneers of this discipline. He has created a memory component forerunner which relies on the protein bacteriorhodopsin, and on laser beam radiation to activate this protein and therefore encode the information.

This prototype device enables the storage of several hundred megabits. The development of a hybrid memory card will increase this capacity to a storage capacity of more than a dozen gigabits.

> Bacteriorhodopsin is a purple colored protein which is found in the membrane of certain bacteria (halobacteria) which are found at the bottom of salty marshes (the origin of the different colors of salt marshes). The bacterium contains a molecular pigment called retinal which is very similar to the one which makes up the retina of our eyes. It is the simplest known protein which is capable of converting luminous energy into chemical energy. Bacteriorhodopsin is therefore a protein for bioenergetics.
>
> The understanding of how this molecule, a real proton pump, functions is achieved by studying its fine structure. Recent work carried out by the Institute of Structural Biology (CEA-CNRS-Joseph Fourier University, Grenoble, France) in collaboration with the University of Basel, has enabled scientists to determine the structure of the molecule.

The rhodopsin molecule changes form when it is exposed to a luminous source and then either works like a switch or a transistor. It can therefore process information. The photo cycle of bacteriorhodopsin is in fact an ideal candidate for storing binary information: the bR-state or logic value 0, and the Q-state or logic value 1 are intermediary structural states of the molecule and can last for several years. This property, which ensures the remarkable stability of the protein, has been created through the process of evolution in the fight for survival in severely hostile regions.

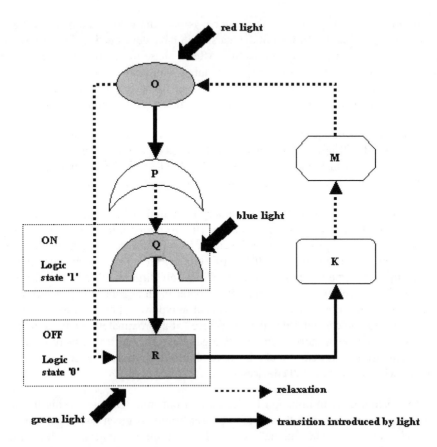

Figure 3.10. *The photo cycle of bacteriorhodopsin or its cycle of molecular states depending on the luminous radiation that it absorbs. The initial state known as bR evolves into state K under the effect of green light. From there it spontaneously passes to state M then to state O (relaxation transitions). Exposure to red light from O state will make the structure of the bacterium evolve to state P then spontaneously to state Q, which is a very stable state. Nevertheless, radiation from blue light can take the molecule from state Q back to the initial state bR. The coding of the logic gates 0 and 1 can therefore be represented by the two most stable states bR and Q and enable the storage of information. It should be pointed out that if state O is not exposed to red light it will automatically pass to state bR*

The advantage of biological materials resides in their very low production costs: the manufacturing of industrial lots is no longer valued at billions of euros or dollars as they are now accessible for only a few thousand euros or dollars.

In effect, by using a simple fermentor it is possible to recreate a salt marsh in which, after a week, the protein in the marsh will have developed sufficiently so that it can be isolated in substantial quantity. After a phase of centrifugation and seven levels of purification, we succeed in obtaining a protein in an aqueous solution which can be conserved from a few weeks up to a few months before being implanted into datacubes which form part of the first organic computer memory.

The fundamental property of a protein-based memory rests in the protein's capacity to take different three-dimensional shapes and organize them into a small cubic matrix structure inside a polymer gel, allowing for the creation of real 3D memories (conventional silicon memories are exclusively two-dimensional).

The storage capacity in 2D optic memory is limited to $1/\lambda^2$ (with λ as the light's wavelength). This represents approximately 10^8 bits per cm². In 3D memory, storage capacity is inversely proportional to the cube of the light's wavelength which is $1/\lambda^3$, bringing the density to between 10^{11} to 10^{13} bits per cm³. The experimental protein-based storage device or datacube suggested by Bob Birge in 1992 was designed to offer a storage capacity of more than 18 gigabytes in an organic storage system with dimensions of 1.6 cm x 1.6 cm x 2 cm. This prototype was, however, well under the maximum storage capacity: this type of device (of approximately 5 cm³) should be, in theory, capable of storing up to 512 gigabytes. The storage density of such a three-dimensional system should be several hundred times higher than storage devices with conventional 2D devices.

The encoding of information is also carried out in a 3D space inside the cube containing the protein with the aid of two laser beams. A green laser beam acts like the address bus of a conventional memory found in our computers. It selects a thin section inside the cube. It is only in this zone stimulated by this green light that it is possible to read or write information. A second layer beam (this time red), coming from an orthogonal direction to the first, activates the protein to store the data.

The prototype turns around a transparent matrix cell called a basin which, by holding the polymer gel, holds the protein. The spatial position of the protein (which is initially at the state bR, i.e. the logic state 0) is supported by the polymer gel. The basin is surrounded by a battery of lasers and a Charge Injection Device (CID) used for encoding and reading the data.

Rebuilding the World Atom by Atom 79

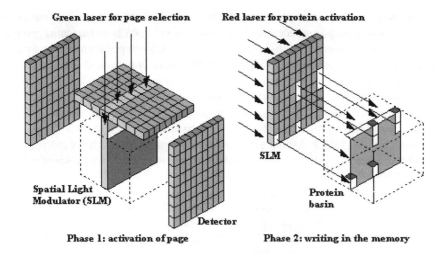

Figure 3.11. *The writing process in a bacteriorhodopsin-based protein memory device*

In order to save data, the green laser for page selection is first of all activated. A spatial light modulator (SLM) which is made up of a liquid crystal device (LCD) operates as a mask on this laser beam. In the matrix, it is possible to only select the precise section that we want to send to the memory. This pre-activated section is the equivalent of a page of data capable of storing a table of 4,096 x 4,096 bits.

After a brief instant, when the majority of illuminated molecules have evolved to state O, the second red laser, perpendicular to a yellow laser, is activated. Hidden by a second SLM, this beam of red light can only reach and activate the zones that we want to put on the logic state 1 on the previously selected section. This flow of red light makes the selected protein molecules move onto state P which spontaneously evolves towards state Q, a stable state which can last for years during the storing of desired information.

In fact, the SLM mask of the red beam enables us to address only points or pixels which need to be irradiated in the pre-selected page by the green laser in order to remain in state Q (i.e. logic state 1). The other zones of this pre-activated page will return to state bR (i.e. logic state 0).

The principle of reading stored information is based on the selective absorption of the red light by the intermediary state O, i.e. the possibility of identifying the binary states 0 and 1 from different absorption ranges according to the structural state in which the protein is found. The process used in writing is also used to simultaneously read the set of bits of data on a page. First of all, a flow of green light lights up the section of proteins to be read which starts the normal photo cycle of the molecules found in state bR. Two milliseconds (ms) later the entire page is subject to a weak flow of red light, sufficiently weak to stop the proteins currently coded 0 in the intermediary state O, from jumping directly to state Q. The molecules which were in state Q with a binary digit of 1 do not absorb this light and do not change state, but those molecules which were in state bR with a binary digit of 0 do absorb these light beams because they have moved into the intermediary state O which absorbs red light. A detector receives the image which crosses the cube, a matrix of dark and light red spots, reproducing the regions in which the molecules in state O and those in state Q are found, distinguishing them as 0 and 1 respectively.

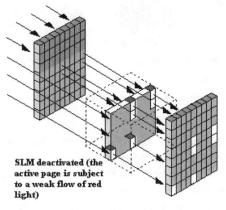

SLM deactivated (the active page is subject to a weak flow of red light)

Phase 3: reading

Figure 3.12. *The reading process in a bacteriorhodopsin-based protein memory base*

To remove data, a short impulse of blue light needs to be applied to make the molecular structure evolve from state Q to state bR. The use of a laser is not needed for a process of mass resetting. The entire basin can be removed in one action with the use of an ordinary ultraviolet lamp. In order to selectively remove pages, a hiding device is used to mask the adjacent pages that are going to be preserved.

The operations of reading and writing use two additional parity bits in order to prevent errors. A page of data can easily be read and reread more than 5,000 times without corrupting the data. In addition, a cycle meter allows for the regeneration of a page by rewriting it every 1,024 reading cycles, making it a sufficiently reliable system to be used in working information systems.

Knowing that a bacteriorhodopsin protein evolves from one structural state to another every 1 ms, the total time of a reading or writing cycle is close to 10 ms, which seems long in comparison to conventional semi-conductor devices. Accessing either the reading or writing process in such a device is quite similar in terms of speed, with rates of up to 10 Mbps. In other words, by lighting a square section with a side of 1,024 bits inside the protein cube, approximately, 105 kilowords can be read or written during the same 10 ms cycle. By combining eight storage cells to make up a byte, the speed reaches 80 Mbps.

The industrial opportunities of this approach, which can be qualified as a hybrid bionic system, i.e. the coming together of living and semi-conductor electronics, still remain far from being achieved.

The transfer of this optic memory bank (of which the core device measures 20 cm x 25 cm) to a small card, from which it is possible to equip an industrial computer, is an easy process. On the contrary, the generic problem of nearly all the new electronic devices which do not stem from semi-conductor technologies is the difficulty of the interconnection of the protein switches which only measure a few nanometers. This problem is far from being resolved, especially the moment we leave the strict confines of the prototype (see section 5.2.6).

Unlike the human brain which integrates several billion cells in the same environment, the applications of this hybrid bionic system, where datacubes come from, are assemblies of protein molecules, semi-conductors and optronic components which create the interface. For the moment, constructing operational machines by using these three types of components in an industrial production process is absolutely unrealistic. Today, only the silicon industry can master the interconnection of a very large number of elementary devices.

> The human brain is the main reference regarding the sophistication and the miniaturization of a universal computer. Adaptable, capable of learning and equipped with the most incredible versatility (survival in any given environment requires the brain to be generalist rather than specialist) it is made up of a precious large-scale integration model with more than 500,000 billion switching devices known as synapses which are heavily interconnected with one another. The brain is a very complex set, made up of networks of neurons with sophisticated and highly performing multiple connections. The neuron receives electrical messages from other neurons. The neuron has a basic logic unit: the cell body which analyzes information and decides whether or not to send a message. If the structure of the brain has been able to resolve the problem of connectivity at a very high degree of cell integration, the history of the evolution of living objects has also achieved some solutions to the problem of mass production. Self-assembled and in constant evolution, the neuron networks can modify themselves in order to adapt to new changes, these networks are known as "plastics". Animals and humans are equipped with the capability of permanently adapting themselves to changes in their environment.

3.6.4. *Spin valves and spintronic semi-conductor components*

Spin electronics or magnetoelectronics is another area of research of nanoelectronics which has experienced extreme interest since the discovery of Giant Magnetoresistance at the end of the 1980s. Current electronics uses only the electron's charge to transport and store information. The use of another fundamental property of electrons, their spin (i.e. their ultra rapid rotation movement on an axis similar to a tiny planet spinning on itself) as a supplementary variable has paved the way for new applications coming from an emerging and promising discipline: spintronics.

> At the end of the 1980s two researchers, Albert Fert from the University of Southern Paris (Université de Paris Sud) in France and Peter Grenberg from the KFA Research Institute in Jülich, Germany both discovered the GMR effect. They observed very significant variations in resistance (6% and 50% respectively) in materials composed of alternating thin layers of different metallic elements. This discovery came as a big surprise to the scientific community whose physicists considered such an effect as theoretically impossible. The experiments had been carried out at a low temperature in the presence of very large magnetic fields and by using materials manufactured in the laboratory which could not be achieved in a process of mass production. However, the dimension of this discovery quickly grabbed the attention of physicists such as engineers since it creates new opportunities for the exploration of a previously unexplored theory, and also for technological applications in the fields of magnetic storage and sensors.

Since 1997, applications of GMR, i.e. a technology founded on the observations of spintronics, have been appearing on the market with the new generation of magnetic hard drives. Other developments such as the Magnetic Random Access Memory (MRAM; see section 5.2.1) should begin to appear on the market in the near future. However, spintronics does not limit itself to only these industrial creations. More innovative components are currently being tested in the laboratory such as the spin transistor, the spin-LED (Light-Emitting Device) or the reprogrammable magneto-logic circuit which will fundamentally broaden the number of applications available in this domain.

In addition to their electric charge and their mass, the electrons are characterized by their spin. This rotation creates a magnetic field (like that of our planet which comes from the magnetic poles). The spin angular momentum can take two directions. Either the rotation is carried out from west to east (with the movement of the Earth) and the spin is referred to as being up, or the rotation is carried out in the other direction (i.e. east to west) and the spin is referred to as being down. In a magnetic field the up and down electrons each have a different energy.

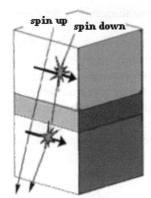 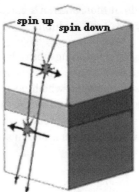

The two magnetic layers are aligned in the direction of the spin: resistance is weak and the electrons flow

The two magnetic layers are not aligned in the direction of the spin: resistance is strong and the electrons have problems in flowing

Figure 3.13. *The GMR effect is generally visible in a multilayered magnetic structure when two layers are separated by a thin space layer a few nanometers thick. The process is similar to an experiment of polarization where two aligned polarizing filters allow light to pass and two intersecting filters stop the light from passing. The first layer enables the passing of electrons whose spin corresponds to one of the states (up or down). If the second magnetic layer is aligned then the canal corresponding to the electrons in the same state is open, in other words, resistance to the flow of electrons is weak (i.e. the current of flow of significant electrons). If the second magnetic layer is not aligned with the first, then resistance to the flow of electrons is strong*

GMR is a simple, yet industrially efficient, process for influencing the spin of electrons and for controlling the electron flow. However, more long term ambitions of GMR aim at manipulating the spin on the scale of each electron with the help of appropriate devices.

The applications of miniaturization such as the execution speed of such devices would go beyond what micro-electronics has let scientists dream of up until now. In fact, our computers exploit the macroscopic flow of electrons (i.e. traditional electric currents) whose intensity allows for the saving, reading, transmitting or processing of binary information.

In a normal electric current, the direction of spin of the electrons is random. This spin does not play any role in determining the resistance of a circuit or the amplification of a transistor. In the world of spintronics each electron (rather than some hundreds of thousands in a current) can, according to the direction of its spin, represent one bit of information.

Therefore, by exploiting not only the charge but also the spin of the electron, great opportunities are being created for information technology. Spin valves or filters can also be created which do not let the current pass if the electron's spin is aligned in the right direction. Such filtering of the current can be carried out on a macroscopic scale by the process of polarization with the help of ferromagnetic materials in multiple layers, such as the current applications of GMR with mass storage.

However, in the longer run, the vocation of spintronics is moving towards unitary control devices such as spin valve transistors.

This new type of semi-conductor uses quantum science and, more precisely, uses a quantum dot, which will be dealt with later in this chapter (see section 3.6.5). The dots behave like artificial atoms, whose charge can be forecast and modified.

By connecting quantum dots to a drain whose spin is polarized by side gates, it is possible to ensure the alignment of the electrons' spin which enters the drain, from either the top or the bottom. By controlling the charge and the electrons' spin at the same time, a type of transistorized quantum processing is introduced (this will be dealt with in the next chapter) whose quantum unit, the qubit, is made up of a precise state of spin.

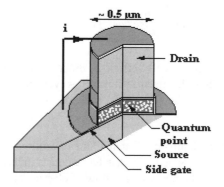

Figure 3.14. *The experimental spin valve transistor is made up of a quantum dot with spin polarized terminals*

Before considering the industrial opportunities for semi-conductors using spin electronics, numerous obstacles have to be overcome such as generating, injecting and transporting a current polarized with spin into a semi-conductor. Once this has been achieved the level of polarization needs to be measured and controlled, and the semi-conductor must no longer rely on the requirement of functioning at a very low temperature in order to avoid the loss of the spin's alignment. If traditional electronics has always ignored the spin of the electron, it is because finding a way of managing the electrons by influencing their spin is not an easy thing to achieve.

3.6.5. *Quantum dots and the phenomenon of quantum confinement*

Semi-conducting quantum dots, or Qdots, are three-dimensional nanometric inclusions of materials with a weak gap (i.e. the property of metallic conductors; see section 2.2.1) in another material. A larger gap matrix characterizes the properties of insulators. The originality of their physical properties and the uses that they can withstand have, for several decades, created much interest from the international scientific community. If they are the bearers of the most technologically advanced industrial opportunities (the essential building blocks for the creation of a quantum computer), they also pose a risk. It is not certain that this research will one day lead to the emergence of a miracle component which would allow for the manipulation of quantum bits and therefore lead to a more concrete vision of such a quantum computer.

In conventional semi-conductor devices, there are numerous free electrons which are responsible for electric current. These electrons, which have varying energy levels, move at different speeds. The dimensions of traditional semi-conductor components, being very large in comparison to that of the atom, mean that the electrons are free to move in the three dimensions of the material's volume. If structures of materials with thin layers are used, reducing the material to a very thin layer means that the electrons can no longer move in three dimensions but in two. If the material is reduced to the size of a single wire then the electrons can only move on this line (one dimension). The ultimate reduction of a material takes the electrons to the confinement of a tiny dot (zero dimension).

In reducing the dimensions of the material, the number of electrons in these quantum dots can also be reduced to a few dozen. When the dimensions of such a dot become ever closer to those of the atom (a strong three-dimensional confinement of electrons), the behavior of the electrons also becomes similar to those found in an atom. In other words, in a quantum dot, just as in isolated atoms, the electrons can only adopt a limited number of distinct states, contrary to materials on a macroscopic scale.

> Approximately 15 years ago, the production of dots whose size was only a few nanometers in which a small number of electrons (less than 100) could be trapped seemed, once again, to come from science fiction.
>
> Quantum dots are, in fact, small piles of atoms in which each pile contains several thousand atoms which spontaneously form when a small quantity of semi-conducting material is placed on the surface of another semi-conducting material with a very different structure. The piles of atoms are created a little like the tiny droplets of water after steam has formed. In addition, the dots have the same shape as the tiny droplets of water but they are a million times smaller and they can only be observed with the help of an electronic microscope. Once formed, they are covered with a protective layer.
>
> These dots can be created at will, meaning that it is possible to modify their composition, their size, their shape and their number during their production. In order to do this, certain parameters such as temperature, the quantity or the speed at which the material is placed on the surface, must be worked with so that their properties (especially optical ones) can be easily changed.

It is predominantly the properties linked to light radiation which make quantum dots interesting.

Light, as a phenomenon of movement, can be seen as a wave or a flow of particles, more specifically known as photons. The emission of light by an atom can be interpreted by the jump of an electron with a larger orbit towards a smaller orbit, potentially less energetic. During this jump to a smaller, less energetic, orbit, the electron gives up a part of its energy in the form of a photon which is emitted towards the exterior.

In reverse, the electron of an atom can absorb a photon with a given energy and jump from a low energy internal orbit to a more energetic external orbit. It is therefore placed in an excited state because it finds itself in a more energetic orbit than the ground state. By naturally de-exciting itself, it will emit another photon.

In a quantum system, energy cannot take arbitrary values: only certain precise energy levels are possible. It is said that the system's energy is quantified. The changes of state also correspond to the precise values of the energy, corresponding to the difference in energy between the final level and the level of origin.

Since the number of states that the electrons can take in the quantum dot is very restricted, the emitted light corresponds to a precise wavelength (i.e. a distinct color). The particular wavelength of the light emitted by quantum dots depends on the nature and size of the dots. This differentiates them from normal sources of light such as our everyday lamps, or natural light coming from the sun, which simultaneously emit a large number of wavelengths. Remember that light appears white because it contains all the visible colors of the spectrum. Not even conventional lasers can emit as pure a light spectrum as the quantum dot does.

The development of more precise and powerful lasers form an ideal application of these tiny dots: by placing a large number of dots side by side, it is hoped that a device that can be used industrially will be developed.

What interests us regarding more precise information systems is the property that these nanodots have of being able to detect quantum information.

Quantum dots are tiny piles (10 to 20 nm) of a certain type of semi-conductor inserted inside another type of semi-conductor. The trapped pairs of electrons and holes which are found there behave almost like artificial atoms. After an adapted excitation (impulse of energy), it is possible to collect restored energy (fluorescent light) by one of these dots and it is also possible to detect, for a given wavelength, an impulse of light which only contains one proton.

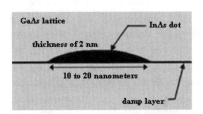

Function of an electronic wave in an excited state inside an InAs/GaAs quantum dot which is epitaxed by molecular jets.

Source IEF UPS

Figure 3.15. *The quantum dots that scientists currently work with are self-made nanostructures which form spontaneously during the crystalline growth of materials presenting a strong grid variance (i.e. singularity), such as InAs and GaAs. The first monolayers of InAs intersect by covering the GaAs substrate (the process of planar epitaxy) and by accumulating considerable amounts of elastic energy. What is known as a damp layer then forms. However, when the critical thickness (approximately two monolayers) is passed, the system minimizes or relaxes its elastic energy by growing three-dimensionally. This leads to the creation of small piles of InAs which push through and create areas of extra thickness in the damp zone. These piles of InAs can be covered with a layer of GaAs in order to achieve quantum dots which resemble hemispheric lenses with a 20 nm diameter and which are 2 nm high*

Quantum dots possessing a resident electron can be produced. This electron comes from the doping of the semi-conducting lattice with electron-giving atoms (for example, silicon atoms that have four electrons on a gallium surface that only has three electrons). Experiments have shown that it is possible through optic methods to position the spin of this electron and to check, through an independent method, the result of this action. In other words, through experimentation it is possible to write and read the spin of a trapped electron in the dot. Carrying out such operations in the values of a quantum variable such as spin (in opposition to traditional variables such as electric charge) forms part of one of the first steps taken towards the actualization of the quantum computer.

If it possible to write and read the spin of a trapped electron in a quantum dot, then we can consider (in spite of production constraints and less so the industrial constraints on a large scale) using these dots as the primary components of a quantum computer, which will be discussed in the next chapter. Let us keep our feet firmly on the ground by introducing these avant-garde devices; using only observed physical effects and tested models do not enable us to create such components in the short or medium period of time, neither do they enable us to produce the practical applications that will use these devices. We are only at the beginning of a scientific epic which may not have any industrial future.

The next stage in this adventure would be the ability to control single dots. For the moment, it is possible to control only groups of dots, which would be an excellent area for experimentation but it is not compatible with the vision of a computer whose miniaturization will come about by the unitary control of quantum dots. Numerous other basic problems will also have to be resolved, for example, the production of interlinked states in two neighboring dots, or even the control by the difference in electrical potential of quantum intrication. Such a number of challenges could, at any moment, either be the end of this latest generation of computers or pave the way for one of the many fabulous opportunities available to information technology.

> Intrication means to tangle. The term quantum intrication refers to the property of every couple (or every set) of quantum objects to appear in a superposition of states. Each of these states describes several objects at the same time, with the properties of these objects all being linked to one another. If, for example, in an object couple named (A, B) object A is in a certain state, this will partly determine the state of object B.

Chapter 4

The Computers of Tomorrow

4.1. From evolution to revolution

When the first computer using transistors was presented to a distinguished committee of experts, their ironic conclusion was as follows: "It is very good. At the end of the 20th century we will be capable of creating 5,000 operations per second with a machine that weighs 1.5 tons and which will consume a dozen kilowatts". This same committee also stated that in the USA, computer requirements would never go beyond a dozen units.

The first electronic computers were considered to be simple laboratory experiments. Since 1995, more PCs have been sold throughout the world than television sets. More than a billion inhabitants on Earth are connected to the Internet. A quarter of the global population has a mobile telephone equipped with a microprocessor which is more powerful than the historical machine mentioned above.

In the span of 40 years, digital computers have changed all economic and industrial activities on a global scale. There has been no similar discovery with such far reaching consequences since that of fire. Not even the automobile, the railway or even the airplane have led to such a transformation in social behavior.

We are, however, only at the end of the genesis of this new digital age. The 21st century will undoubtedly develop as the maturity period of this technology. Its penetration in all aspects of our daily lives will perhaps be as unbelievable as was the development and spread of the first prototype of the transistor computer.

The entry of information technology to the 21st century is, above all, the history of a non-event: that of the regular development and rapid progression of processors and platforms stemming from the traditional silicon industry, which is an industry that works on successive generations of smaller transistors for faster functioning devices.

This industry will inevitably end by reaching its technological limits within a scale of 10 to 20 years. It will continue to provide the essential tools necessary for information systems during what is a sufficiently long period in the eyes of an economist or in the eyes of market pragmatism. As Keynes once said, "in the long run, we will all be dead".

The improvement in the performance of computers will most certainly continue at its current exponential rate for at least a decade. Apart from any ground-breaking innovation beyond this period, the only progress in this field will be limited to minor perfections of a technology which has achieved its asymptotic performance.

This means that the progression of this technology will be similar to that of other technologies which have arrived at maturity. At most, it will evolve only a few percent each year (take the example of thermal engines: the efficiency of internal combustion engines and turbines use basic technology which has remained the same for over 100 years and progress in a very incremental fashion). The example just given has nothing in common with the silicon industry, so in order to regain the rate of progress to which we have become accustomed, something else must be discovered.

What will computers be like throughout the next decade? The machines that are sold on the market are not upgraded and replaced from one day to the next. Just like the slide rule, analog computers and the first digital computers have co-existed for more than two decades. Ground-breaking technology will progressively emerge in a market largely dominated by machines that are a continuation of current basic technology.

This is why it is interesting to consider two visions of the future of the computing industry for the next decade. These two visions oppose and complement one another:

– the first is the usual techno-economic vision or incremental evolution with a term of one to three or even five years. This is the idea which conditions traditional industrial investments and which leads to the effective growth of the economy;

– the second vision is the anticipation of technological limits which should be reached only some years beyond the fixed term which is set by traditional market rules. This anticipation recommends evaluating the characteristics of a new generation of computers with radically different structures and which could come into being in approximately 10 years' time.

Between these two eras in informatics, an intermediary period exists. This is a period which is still quite vague for analysts and technologists. The most cautious amongst them predict that the silicon industry will continue to impose its traditional platforms with technology that will achieve such remarkable performance which will satisfy the needs of the most ambitious projects and systems.

However, research and the first applications in the field of nanotechnology and new disciplines in applied physics predict a progressive incursion of other solutions.

Certain functions will exploit the devices which are in competition with conventional semi-conductor components. The problems linked to cryptography or organization and scheduling could thus be processed by specific specialized processors produced from non-silicon technology.

Large capacity memory and storage will exploit radical new methods similar to those used in holography, or they will exploit new methods which would be taken from organic molecular chemistry. The analysis and diagnosis of internal diseases could rely more and more on the use of biochips.

These dedicated sub-systems with a performance superior to those of their rivals from the traditional silicon industry will co-exist in environments in which silicon technology will continue to carry out the large part of the functions.

For example, the resolution of specific and complex problems regarding industrial organization or logistics could be processed by a specialized and dedicated, perhaps non-Boolean, high performing device. This device would work in collaboration with a conventional system such as ERP (Enterprise Resource Planning). These applications offer inexhaustible opportunities.

92 Nanocomputers and Swarm Intelligence

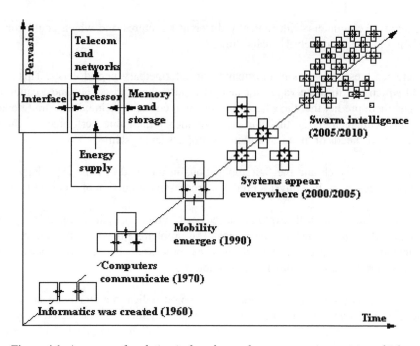

Figure 4.1. *A process of evolution in five phases: from a processing activity, which was exclusively computer-centric, to swarm intelligence. It should be pointed out that all mobile and communicative systems, whether they are conventional or tiny and atomized, function schematically around five components: (1) the processor, (2) the memory and mass storage devices, (3) the interface between man and machine, machine and machine or environment and machine, (4) the telecommunication infrastructure and the networks which enable systems to communicate with one another, (5) the energy supply which becomes an integral part of mobile and autonomous devices*

However, by continually pushing back the limits of technology it is the concept of the information system itself which is changing: from a processing activity which was centered exclusively on the computer, informatics and the world of computing are becoming swarm intelligence. As technology allows for the production of tiny computers which are available everywhere, it means that almost all of our everyday objects can activate a spontaneous exchange of information without the interaction of their user.

It is a fundamentally different model which will introduce profound changes in the business world and in our daily lives, and which will inevitably bring about a major evolution in the computing world in terms of activities and jobs carried out in this field.

> A biochip is a tiny analysis and diagnostic tool which is approximately 1 cm². It appears in the form of a glass or silicon support on which proteins or millions of fragments of DNA or RNA are fixed. For a relatively low unitary cost, these biochips enable a rapid identification of a large number of genes (DNA or RNA chips) and of proteins (protein chips). These biochips also enable scientists to study how DNA and RNA genes and proteins function. They are also part of the precursors of what could become the pillars of the future.
>
> DNA and RNA chips facilitate the diagnosis of infectious or genetic diseases. They contribute to the research in resistance to antibiotics from bacterial strains. They help analyze genetic mutations and identify new therapeutic targets. They also allow for a better understanding of how medicines work by identifying their secondary effects. Their use in specific treatments according to the genetic profile of the patient is also being considered. DNA chips also allow for a relatively easy and cheap traceability of food products, the detection of GMOs, or even the bacterial analysis of water. Protein chips have more recently led to interesting applications in the development of new medicines.
>
> The principle of the DNA chip rests on the particularity of spontaneously reforming the double helix of DNA opposite its complementary strand. The four base molecules of DNA are unique because they join together in pairs, adenine with thymine and cytosine with guanine. If a patient is a carrier of a disease, the strands of DNA extracted from this patient will hybridize with the synthetic strands of DNA which represent this disease.
>
> In order to improve research in genes and more complex forms, researchers have teamed up with electricians in order to exploit the properties of silicon chips. Companies from the biomedical sector (for example Affymetrix, the American leader in acquisition and analysis technologies, and in the management of complex genetic information) currently use the photolithography techniques which are only used in the microelectronics industry.
>
> DNA matrices and biochips are mainly used in genetic research, in research in pathology and in the development of medicines. The PolymerMicroSensorFab project supported by the nanotechnology initiative of the European Commission aims at developing a disposable biochip for DNA analysis at a very low unitary cost.

4.2. Silicon processors – the adventure continues

The leading companies in the silicon industry have already taken their first steps into in the field of nanotechnology. This started at the end of 2002 when the sub-100 nm barrier was broken. For example, Intel, which is one of the pioneers of the silicon industry, views nanotechnology as a continuity solution from conventional silicon. It is true that certain groups such as Intel or IBM have become masters in the art of building small objects by investing heavily in this field. They undoubtedly make up the leading private budgets in the area of nanotechnology.

It is also true that these leading companies, whose business is totally dependent on a technology that they do not want to see decline, will defend their nest egg tooth and nail. It was the same for petroleum engineers regarding alternatives to the internal combustion engine.

The silicon industry is fundamentally performing over industry average. In 2000 the majority of analysts considered that the average growth of 15% per year, as was experienced by the semi-conductor industry between 1960 and 1995, was definitely in the past. They estimated that it could not hope to do better than 7% to 10% per year throughout the decade. Despite all expectations, in 2004 there was a 19% increase in the global sale of electronic chips.

The industry analysts expected a pause during the last few years that would absorb the rapid growth of the silicon industry. The launch rate of new technological products has gone from every three years before 1995 to merely every two years ever since. The analysts and experts thought that there was a lack of these famous killing applications which are essential for boosting growth. These killing applications include emerging platforms and especially the wireless sector, and it is these applications which have largely maintained this market growth. General public markets on a global scale have strongly contributed to this growth at a time when consumers are adopting new technology and purchasing multifunction digital devices such as mobile telephones equipped with cameras, personal digital assistants, DVD players or recorders, and flat screen televisions. The delivery of PCs has also contributed to this growth by recording an increase of 11% in 2003 which translates to more than one billion Internet users two years later.

65 nm technology arrived in the market in 2005 at Intel as well as at AMD who signed a development agreement with IBM, and 45 nm technology are likely to appear from 2007 onwards. The silicon industry still has a bright future even if there are very few manufacturers who produce silicon wafers (a factory which is capable of creating 30,000 silicon wafers per month represents an investment of more than $3 billion). The obstacles faced by the next generations of semi-conductors will be more economical than technical.

> Launched in 2005, the processor which bears the code name Smithfield ushered in the era of dual-core processors at Intel. From 2006 onwards, middle of the range PCs have been equipped with dual-core processors with an Intel processor bearing the code name Presler. The Presler, which has a 140 mm² chip and is manufactured using 65 nm technology, is a shrunken version of the Smithfield, Intel's first dual-core processor for desktops.
>
> Even though the advent of dual-core processors undoubtedly represents progress for desktop PCs, operating systems and applications software will have to be reformed in order to take advantage of the excess in processing power. This is a prerequisite condition because the current applications software and operating systems which are developed for only one thread work much slower on dual-core processors than on their single-core counterparts.

4.2.1. *Progress in photolithography and new materials*

Photolithography is the process used for producing the circuits and all other electronic devices which are found in our computers. The tools used in engraving by lithography work just like the principle of traditional photography (see section 2.2.3). It is a complex process which uses a luminous source which is focused and then directed with the help of lenses and mirrors. This light is then filtered with the help of a photo mask to insolate only the image to be engraved on the silicon wafer.

The need to create finer structures has led to an evolution of the traditional techniques of photolithography allowing for the transfer of patterns with a size of less than 100 nm. Different evolutions in photolithography should enable the limits of the thinness of silicon engraving to be pushed back further.

Electron-beam lithography was the first technique that made sub-100 nm production possible. Today this process also allows for the best resolution at a size of 10 nm thanks mainly to the use of focused electron beams. Ion-beam lithography is a method of direct engraving which uses a source of ions and an ion optic to focus the beam. It has several advantages over electron beam lithography, mainly a more significant ion mass. However, this variant of lithography still does not perform as well in terms of resolution. One of its evolutionary processes, lithography by ion projection with or without a mask, is limited to an engraving thinness of 70 nm.

X-ray lithography uses X-rays with a wavelength of between 0.5 nm and 4 nm enabling it to have both a good resolution (of less than 50 nm) and a good depth of field (penetration). It remains a delicate technique due to the difficulty of efficiently concentrating this type of radiation.

Extreme ultraviolet (EUV) technology is quite similar to the current traditional processes of lithography, but in terms of the thinness of the engravings EUV technology will offer a much superior quality than what is available today by replacing the objectives (transmission masks) with a series of precision mirrors (reflection masks) which will make it possible to focus UV radiation with a wavelength of 10 to 15 nm. It will also lead to a smaller resolution at 45 nm. This type of technology is more than likely the best candidate for the industrial development of engravings which are less than 100 nm in size. When this technology arrives at maturity it will probably lead to the ultimate CMOS technology (sub-30 nm).

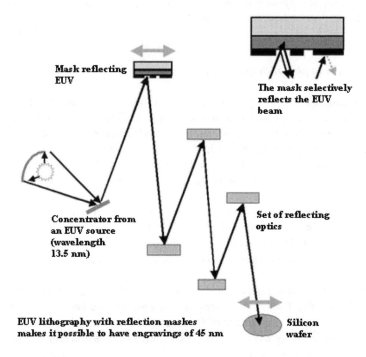

Figure 4.2. *The process of EUV photolithography authorizes a thinness of engraving superior to that resulting from conventional processes by replacing the current objectives (transmission masks) with a series of precision mirrors (reflection masks). This technique makes it possible to focus UV radiation with a wavelength of 10 to 15 nm*

Other processes which make it possible to create nanostructures on silicon chips with an increased level of thinness also exist. The process of soft lithography consists of creating an elastic stamp which enables the transfer of patterns of only a few nanometers in size onto glass, silicon or polymer surfaces. With this type of technique it is possible to produce devices which can be used in the areas of optic telecommunications or biomedical research. However, this process still cannot be used for the industrial production of electronic components. In fact, the patterns are obtained by the process of stamping thanks to a shape (or stamp). This stamp is produced from the normal techniques of photolithography or electron-beam technology. The resolution depends on the mask used and can reach 6 nm.

This process is not a direct candidate for photolithography in the creation of silicon components. However, the properties of self-assembly on a nanometric scale which this technology uses makes it possible to create micro-actuators that can be used in optronic components for the equipment in fiber optic networks and in the long term for future optical computers. It is also a very valued technique which is used in the creation of labs-on-a-chip for the direct printing of proteins. The PDMS (polydimethylsiloxane) stamp is used to transfer and structure the proteins on a surface which becomes covered with a network of cylindrical pins made up of elastomere in the nematic state (i.e. the mesomorphic state which is a state that is closer to the liquid state than the crystalline state). These pins suddenly contract by 30% to 40% to the nematic/isotropic transition state of the elastomere, and they act as micro-actuators in order to separate and confine the DNA.

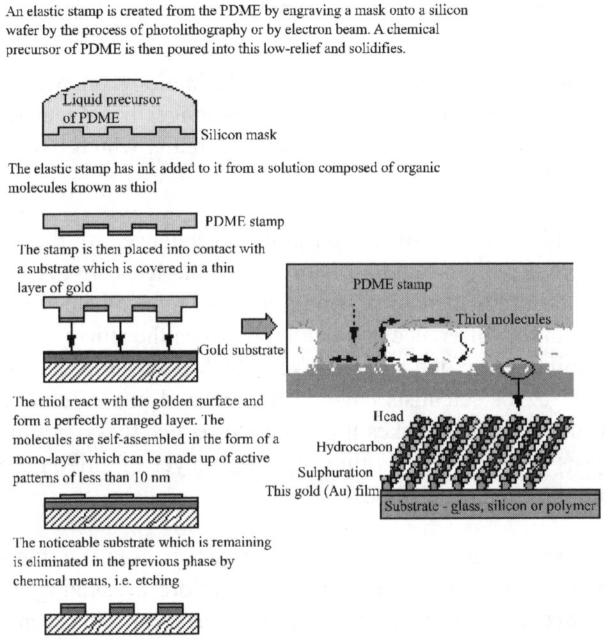

Figure 4.3. *The process of soft lithography consists of creating an elastic stamp which allows for the transfer of tiny patterns of several nanometers in size onto glass, silicon or polymer surfaces*

> Since the beginning of 2004, with a common engraving thinness of sub-100 nm, the problems that exist no longer only affect the processes of lithography. New types of constraints are emerging, notably in the interconnection of the more numerous, more miniscule and denser transistors. These constraints are also found in the internal functioning of the transistors themselves. With engravings of 65 nm, the time has come for the current silicon dioxide to be replaced. With engravings of 90 nm, the gate dielectrics (insulator) must not have a thickness greater than four atoms if it is to be possible to reduce and master the internal capacitance, a parasite capacitance which has to remain compatible with the desired switching speeds. Quantum effects called tunnels appear and the electrons start to flow to where they are not wanted. This is where the notion of mutual disturbances between neighboring wires comes from. Thicker dielectric materials with a strong transfer rate are needed in order to reduce these leaks and at the same time conserve electrical capacity. Amongst the candidates for replacing silicon dioxide there are aluminum trioxide, hafnium dioxide, and zirconium dioxide. New dielectric materials with a weak transfer rate are also needed for the insulating layers at the interconnection level. For example, silicon dioxide doped in fluorine or carbon is already used for engravings of 130 nm and 90 nm. However, certain materials break easily. Careful attention must be paid to the expansion co-efficient in terms of the heat and positioning of the elements, this poses a new development problem for industrialists in the silicon industry.

CMOS technology (see section 2.2.2), the technology on which current chips rely, seems to have reached its limits. It can no longer be executed in engravings which are less than 100 nm, i.e. state of the art circuits which have been available since 2004.

The main problem with CMOS rests in its capacitance, i.e. the internal parasite which slows down the speed of a transistor's switching capacity. At the periphery of the source and the drain there is a strong accumulation of electrical charge. This charge comes from the difference in potential between the silicon substrate and parts of the ionized epilayer (i.e. the surface layer which is close to the source and drain zones). Since the 1990s scientists have been interested in a solution known as SOI (Silicon On Insulator) which makes it possible to reduce these harmful effects. This approach consists of interposing a thin insulating layer of silicon dioxide between the epilayer and the silicon substrate.

In order to form an epilayer, the epitaxy process is used (i.e. the process which enables atoms to place themselves correctly, therefore organizing themselves in a continuous monocrystal with no defaults at the interfaces). This seems to be, at first glance, a simple solution, however, its implementation is delicate because the growth process in epitaxy signifies the continuity of the silicon substrate. It is therefore impossible to insert the insulator by conventional methods after this operation.

The idea is therefore to create the insulator (silicon dioxide) inside the silicon by using a technique known as SIMOX (Separation by IMplantation of OXygen). SIMOX involves an injection of purified oxygen at a very high temperature into the silicon wafer. The injection is controlled in order to achieve layers of oxygen of 0.15 μm. The oxygen which is introduced at a high temperature sticks to the silicon and thus forms thin layers of oxide.

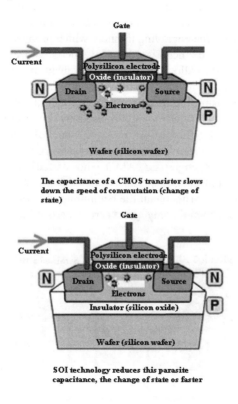

Figure 4.4. *The SOI (Silicon On Insulator) process inserts a thin insulating layer of silicon oxide between the epilayer and the silicon substrate. This solution makes it possible to reduce the internal parasite capacity which slows down the transistor's switching capacity*

SOI technology is often combined with the copper metallization process. Copper possesses a better electrical conductivity than aluminum, and the association of such a metallization process with SOI technology enables the new circuits to function at much higher frequencies than in CMOS technology.

The quest in the search for new materials accompanies more than ever the incremental development of photolithography in the art of building small electronics. Moore has always considered that the continuation of this evolution at the rate stated in his law would force joint progress in the development of new emerging techniques and in the research of new materials.

Constructing ever smaller circuits which work faster in terms of electronic speed signifies the elimination of all obstacles from an electron's path. In 2003 IBM and Intel came up with another improvement which has developed and become widespread since then. It is an approach which works with the same material structure and with the same engraving thinness which produces processors that work up to 20% faster. Strained silicon is a layer of silicon whose crystalline network of atoms has been strained to allow for a faster movement of electrons (by spacing the atoms the interference is reduced). This gives rhythm to processors with higher frequencies. By increasing the distances between the silicon atoms the processors consume less energy. This technology is used by Intel and IBM, especially in Pentium 4E (Prescott 0.09 micron). At the same time, IBM and Intel had announced that strained silicon would produce processors that work up to 20% times faster for an equal thinness of the engraving. AMD who started to incorporate a form of strained silicon in its processor products with 90 nm technology announced a different approach to this principle at the beginning of 2004: a more localized form of strained silicon which would only affect certain parts of the chip.

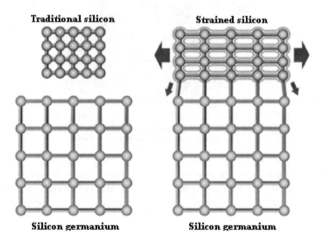

Figure 4.5. *Strained silicon technology makes it possible to reduce the obstacles in an electron's path. With an equal thinness of the engraving it is possible to produce faster processors which consume less energy*

4.2.2. *The structure of microprocessors*

A computer is a complex set of hardware and software designed as a hierarchy or pile of growing abstraction layers. Each abstraction layer has its own language made up of instructions. A lower layer (n) translates or interprets the instructions of the upper layer ($n+1$) by using its own instructions or services. Likewise, each lower layer (n) provides services to the upper layer ($n+1$).

The structure of a microprocessor defines its characteristics such as those detected by programs which run on this circuit. Such an architecture specifies the set of instructions that the processor knows to execute, the number and organization of data registers (i.e. the work storage zones during the execution of an operation) and the configuration of the bus entry/exit (i.e. the physical canals through which data is transmitted from one entity of the computer to another; each bus is made up of two parts: the data bus on which data passes through, and the address bus which makes it possible to direct the data to its destination).

When this architecture is defined, the processor's set of characteristics is irreparably constrained (in fact, this clarifies the programming language which corresponds to the hardware structure where each bit of instruction corresponds to a hardware device such as logic gates). Any unknown quantity in this structure would lead to the incapacity of correctly running certain higher-level programs. It is for this reason that the structure of computers sold on the market evolves very slowly over time and that very few fundamentally new structures are commercialized.

Today, four types of structure[1] which characterize microprocessors are distinguished. The first one is the CISC (Complex Instruction Set Computer) structure which constitutes the basis of all x86 type or Intel compatible processors. RISC (Reduced Instruction Set Computer) technology is the next type of structure followed by the VLIW (Very Long Instruction Word) structure, and finally a more recent structure called EPIC (Explicitly Parallel Instruction Computing). It should be pointed out that certain market products can combine several of these structural models.

The CISC structure makes up the historical approach of the conception of the microprocessor. It remains present in the x86 series (i.e. processors mainly manufactured by Intel, AMD or Cyrix) which was created with the Intel 8086 in 1978. Remember that at this time system memory was an expensive, slow and rare commodity (the largest systems only had several Mbytes of memory and the

---

[1] This typology according to four categories of processors is the current vision which is accepted by the market. Sometimes, the categories are also classified into three: CISC, RISC and VLIW. EPIC is considered as the successor to RISC.

ancestors of the PC only had several Kilobytes of memory). CISC had the authority to minimize the use of this resource. Furthermore, the number of transistors available on a circuit at that time cannot be compared to today's figures. The Intel 8086 had barely 29,000 transistors (the Pentium 4 HT has nearly 6,000 times more with 167 million; see section 2.3.2). Intel therefore opted for a compromise. They made a decision which 25 years later is still greatly impacting the functioning of this long line of computers. They made the decision to use a small number of registers and to use very complex instructions of varying length which require a significant number of clock cycles to be executed by the computer. Processors using a CISC structure (these also include VAX processors by Digital Equipment Corporation or the 390 series by IBM) can process complex instructions which are directly wired or engraved onto their electric circuits, i.e. certain instructions which are difficult to create from a microprogram of basic instructions are directly printed on the silicon of the chip and can be called macro commands in order to increase the rate of execution (otherwise everything else is equal, in the technological context dating from the end of the 1970s).

> On a physical level, a microprocessor uses a very limited number of microprogrammed instructions (AND logic gates, OR logic gates, the loading of a register) which are engraved directly onto the silicon.
>
> Each machine instruction which is supplied to the microprocessor is executed as a series of microprogrammed instructions.
>
> It is impossible to engrave all machine instructions on one processor. The processor therefore elaborates on the machine instructions which make up the exchange language with the microprocessor as a series of microprogrammed instructions.

Unlike the CISC structure, a processor which is based on RISC technology does not have supplementary wired functions. The structure, which was designed by John Cocke at the end of the 1960s, recommends the use of simple and short instructions with identical lengths making it possible for the instructions to be executed very rapidly, i.e. in only one clock cycle. The instructions are synchronized inside a pipeline. A pipeline functions like an assembly unit in a factory or like a musician who on each beat plays a note from his piece: each beat corresponds to a very simple basic task that needs to be accomplished, such as looking for an instruction in the register, decoding it, i.e. interpreting what must be done, executing it then storing the result (modern processors divide these operations into more than four stages). Certain super pipelines of Pentium processors work in 14 stages.

Since the execution time of these instructions is identical, it is possible to execute them in parallel by grouping several of these pipeline processors in a global pipeline structure. We are talking about superscalar machines. An elementary pipeline has at least two levels and works in four stages. The processor simultaneously executes several instructions during each cycle time. Each level can communicate its result to the other at the end of the cycle. In this way, programs can be executed faster. It should be pointed out that the processor can reorganize the instructions as it likes during the process in order to optimize the sequence: the processor examines the flow of the incoming instruction, decides the most optimal way possible to reorganize the set of instructions by taking into account some simple rules, and then checks that the new sequence does not depend on the result of a preceding instruction. This is what is known as dynamic execution, which makes modern processors very complex.

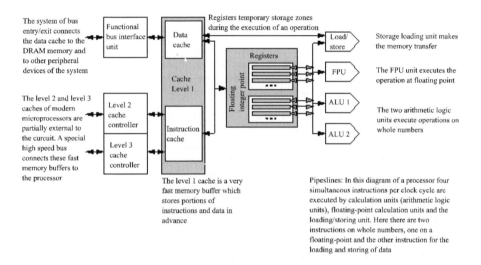

Figure 4.6. *The typical structure of a microprocessor: arithmetic and logic units, registers, pipelines, caches and buses*

However, the most advanced processors go even further in this optimization by using a process of operation prediction of the program. When this program meets a conditional connection, i.e. an if – then – else algorithm, the processor predicts the route that the program is more than likely to take and thus organizes the instruction sequence in this direction. Such a prediction is generally simple and is usually based on the history of sequences. If the processor has made the right choice, i.e. what occurs statistically in 80% to 90% of cases, it will have already anticipated certain sequences of instructions and therefore have gained processing time in relation to common operations. If the processor has made a mistake it will have to retrace its steps by emptying the parallel pipeline of the incorrect sequence of instructions and restore the parallel pipeline with new instructions.

The penalty of such an error is time wasted in the clock cycle. This is made worse because of the long length of the pipeline used. Generally, processors that use longer pipelines can also function at higher frequencies because the longer the pipeline is the more operations that can be divided, i.e. the instructions become much simpler. This is another form of compromise that the designers of microprocessors have to arbitrate.

> Modern microprocessors work at speeds of several GHz, which means that the processor looks for a new instruction every fraction of a nanosecond. However, the DRAM memory which contains the code of the program to be executed requires more than 50 nanoseconds to extract instructions from memory cells and to transfer the instructions to the processor.
>
> To resolve this problem cache memories are used. Cache memories are blocks of very fast memory which, like buffers, are placed between the processor and the system memory and store portions of instructions and data in advance in order to make them quickly available to the processor.
>
> Current microprocessors have two cache levels: the primary cache (or L1 for Level 1) which is generally engraved on the circuit of the microprocessor, and the secondary and tertiary caches (L2 and L3 respectively) which are partly installed on the circuit but whose capacity can be increased by a circuit which is external to the processor. When the processor needs an instruction or a piece of data, it will first of all search the L1 cache (which is the fastest, and in the case of error it will search the L2 cache followed by the L3 cache which generally have a better capacity).
>
> It should be pointed out that caches represent the largest number of transistors which a microprocessor possesses today. The Pentium 4 HT, equipped with 167 million transistors, uses more than half of its transistors to form its caches. The Itanium² Madison 9M processor with its 600 million transistors is equipped with an L3 cache on its circuit, bringing its capacity to 9 Mb. In order to make the distinction between the processor and the cache memory on the chip of a microprocessor, the part of the circuit which makes up the heart of the processor is called the core, and the memory buffers which were mentioned earlier are referred to as caches.

EPIC uses the parallel processing of data like superscalar machines do. 64 bit microprocessors from Intel's new Itanium generation have developed the highly performing EPIC structure from ILP (Instruction Level Parallelism), which was jointly initiated and developed by HP and Intel.

Unlike RISC technology in which the processor organizes the sequences of instructions, in an EPIC model it is the compiler that is in charge of optimizing the code in order to benefit from parallel processing.

In the prediction of connection probability carried out by RISC processors, it is the statistical behavior that conditions the decisions to direct instructions to be processed to the correct pipeline. The error rate of 10% to 20% which penalizes the processing time of a program can be minimized if optimization is carried out from an algorithm which is run by the compiler. The compiler takes context into account rather than acting on the basis of past behaviors.

EPIC processors evaluate the different possibilities in parallel with the correct possibility being conserved. For this reason, processors separately manage what is known as a significant flow of data (a process making it possible to optimize the size of data processing) in order to manage all computations in parallel without any loss in speed. EPIC technology gives the compiler the dominating role as well as offering it new opportunities to find parallelism amongst instructions.

In many applications, and in particular in multimedia applications which create a large number of surprisingly simple basic operations, the key to the performance of microprocessors resides in what is known as fine grained parallelism, i.e. the parallel execution of the most basic instructions.

The principles of pipelines, multiple processors and superscalar machines used in RISC structures function in this way. The concept of VLIW introduces a fourth idea: regrouping several basic independent operations within the same instruction word which will be read then executed by the processor in one cycle time.

The VLIW structure relies on a more complex set of instructions than those of the CISC structure. While RISC technology consists of simplifying the set of instructions by shortening them, VLIW takes the opposing view by encoding four instructions (or more) in the same instruction word, i.e. in one machine operation. The VLIW structure is used in Hewlett Packard PA-RISC microprocessors and also forms part of the structure of Intel's Itanium processors.

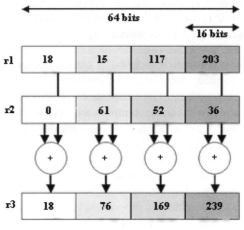

Figure 4.7. *Multimedia applications require more processing power for tasks such as the compression and decompression of sound and video in real time. Some current processors contain specialized instructions known as multimedia which increase the processor's performance. Multimedia applications ordinarily manage values of 8, 12 or 16 bits, which wastes the resources of current processors capable of working on words of 32 or 64 bits. Multimedia instructions get around this problem by simultaneously processing several of these short numbers (i.e. 8 or 16 bits) in larger registers that are available. This can be seen as a type of vector processing with, typically, four to eight components. This vector-type parallelism is the origin of the name Single Instruction Multiple Data. Today, compilers are still not capable of effectively managing such instructions and they involve the manual writing of the program in assembler code, unless the program is restricted to a library of predefined but limited routines. In any case, the search for parts of code, which can take advantage of these instructions, has to be ensured by the programmer*

Since the mid-1990s, the manufacturing of microprocessors has progressed to the point where it is possible to integrate complete structures from this type of computer onto a single circuit. The forthcoming years should see the spread of the PC-on-a-chip, i.e. chips which integrate a microprocessor, a graphics circuit and a memory control all on the same silicon chip. Such a process may reduce certain bottlenecks between the heart of the processor and the sub-system (i.e. bandwidth improvement), and may also reduce production costs. Since the size of the chip is limited, the functions which have been added to the microprocessor's circuit generally do not perform as well as systems which have an external circuit with a chip. These sub-microcomputers primarily concern mobile applications (such as smart phones, PDAs, etc.) or certain entry-level platforms (see section 4.3.2).

4.2.3. *Digital signal processing and DSP processors*

The majority of electrical signals which conventional computers manipulate are binary or logic values. These are signals which only have two states: go and stop. Processors have been historically designed and optimized to manipulate such values.

The greatness of physics in the real world (the speed of a vehicle, the temperature of an oven, the volume of sound of an acoustic musical instrument, etc.) evolves, however, according to a continuous spectrum of values. The series of discrete values that represents the translation of these analog values into a cascade of binary information which can be understood by a digital computer is an abstract of reality which is made all the more difficult because scientists wish for an accurate, reliable model.

For the specialized processing of certain leading analog devices, it is possibly more efficient to use semi-conductor devices capable of directly manipulating the analog signals rather than converting them into binary values. However, such equipment is very expensive. By not using binary digit logic, these analog devices do not benefit from the exponential progress as illustrated by Moore's Law.

Digital Signal Processors (DSPs) are part of the compromise. Specialized in the real-time processing of leading analog devices found in the real world (after being previously converted into a binary digit form), these processors with an optimized structure can carry out complex calculations in only one clock cycle. This specialization also enables them to easily access a large number of digital, as well as analog, entries and exits.

Today, DSPs are largely used in the majority of real-time DSPs. They are also found in modems, mobile telephones, audio and video equipment, i.e. each time a complex physical signal is acquired which is then processed by a filtering operation. In terms of price and consumption, DSPs perform much better than a solution that is based on generalist microprocessors which need a much more logical series of programmed instructions (the idea of physical acceleration since it is obtained by the physical wiring of the calculations to be carried out in the form of a set of specialized instructions where only one of these instructions is capable of carrying out a complete processing operation, unlike the software method where a conventional processor follows a program of generalist instructions).

The basic function used in DSPs is the MAC (Multiply and Accumulate) function, i.e. a multiplication followed by an addition and then the result is stored. This function is the key to nearly all servo-control and filter calculations. DSPs operate scalar products or convolution products[2] as well as acquisition and restoration operations during one clock cycle.

Today DSP manufacturers are putting more and more instructions onto their circuits, and the number of instructions on the circuits corresponds to the number of complete operations of signal processing. Since the end of the 1990s these circuits have been using an increased level of parallelism, a more significant quantity of memory on the chip (like the caches of conventional microprocessors) and a concept of VLIW which is similar to that used by Intel in its new generation of 64 bit processors. For example, the TMS320664 DSP processor from Texas Instruments engraved with 90 nm technology and with a clock speed of 1.1 GHz can carry out nine billion MAC operations per second, i.e. 10 times more than the previous generation's DSP processors.

Since the 1990s it has been communication equipment and, in particular, equipment used in voice processing which has benefited from DSP technology. The growth of the wireless telephone industry, mobile telephones and Voice over Internet Protocol (VoIP) applications will, in effect, boost the demand for this type of processor throughout the decade.

The next generation of objects used in communication capable of interacting with the physical world should only reinforce this appetence for DSPs.

> In 1978 Intel launched its first analog signal processor, the 2920. It was a circuit which had an analog/digital converter (ADC), a digital/analog converter (DAC) and an internal signal processor all on the same chip. However, the circuit did not have a wired multiplier on it and, as a result, was not competitive on the market. In 1979 AMI launched the S2811 which was designed as a peripheral processor and had to be initialized by its host. It, too, did not experience the success which was expected of it.
>
> The same year Bell Labs presented the MAC 4 microprocessor which was a monocircuit DSP precursor. It was not until the following year that the first real DSPs arrived on the market: the μPD7720 from NEC and the DSP1 from AT&T, which were both presented at the IEEE's International Solid-State Circuits Conference in 1980. The two processors were largely inspired from research carried out in the telephone industry. The μPD7720 processor was a fixed point DSP with a clock speed of 4 MHz, a RAM of 128 words, a ROM of 512 words and a program memory of 512 words. It had a VLIW-type instruction format enabling it to simultaneously carry out all types of arithmetic and logic operations, as well

---

2 Convolution is the most fundamental operation of signal processing. It indicates that the exit value of a signal at a moment t can be obtained by the (integral) summation of the past values of the excitation signal $x(t)$ which is balanced by the impulsive response $f(t)$.

> as the incrementation and decrementation of the register and the loading of the operation in only one clock cycle.
>
> In 1983 Texas Instruments presented the TMS32010 which would prove that DSP circuits could be part of true commercial success. At the moment, Texas Instruments is the leader in the field of generalist DSP circuits.

4.3. Conventional generation platforms

Today the market of computing platforms is divided into six different categories, with five of them representing the classic categories: desktops, laptops, workstations, servers, mainframes and supercomputers. The boundaries between these categories are, however, becoming more and more blurred. Middle of the range servers are increasingly inheriting the characteristics of mainframes, such as separate partitions allowing for the simultaneous execution of multiple versions of the operating system. New servers with 64 bit processors are starting to penetrate the exclusive domain of supercomputers. Supercomputers have created a new generation of computers which uses standard components from the market and, in doing so, they have adapted themselves so that they can also be used in business, such as data warehousing on large databases or any sort of processing in marketing which requires large processing power.

The sixth category appeared at the end of the 1990s almost without anyone realizing. This category includes emerging platforms which are increasingly becoming part of our daily lives. Whether it is PDAs or electronic diaries, mobile telephones (which today have the largest number of transistors), video game consoles or navigation systems in cars, these new tools have undoubtedly become one of the major shifts in the growth of digital storage devices. These devices, which are extremely specialized and which have remarkable levels of performance, have rapidly become present in areas that were once thought to be exclusively reserved to traditional platforms. This is the case for the transfer of funds from one bank account to another by telephone, purchasing items online, and accessing the Internet from a PDA navigator or from any other intelligent device such as a digital camera or a mobile telephone. This type of functionality is starting to penetrate the world of household products such as being able to access the Internet from a fridge or a washing machine.

These platforms have even shaken areas of the computing world where they were least expected to. At the end of 1999, Sega, in partnership with Nomura Securities, gave users of its Dreamcast games console the opportunity of carrying out stock-exchange operations on the Japanese market. The most serious or sizeable transaction could be initiated through using what was still considered as a simple toy. At the same time Sony, with its PlayStation 2, provided its users with the performance of a supercomputer which had a processing capacity of more than 6.2 FLOPS (millions of floating point operations per second), i.e. a processing power equivalent to that of top of the range workstations that are normally reserved for the computer animation studios.

4.3.1. *Traditional platforms*

It is under the term "traditional platforms from a conventional generation" that the heritage of current data processing and the large part of the commercial market supply of information systems (without omitting a 5 to 10 year market perspective) is organized nowadays. In other words, in the medium term, even future technology cannot ignore the products that will continue to form the base of information systems over the next 10 years.

These traditional platforms, which range from the domestic PC to workstations, servers and other mainframes, and supercomputers, all have a clear, well established roadmap, i.e. the clear, foreseeable global evolution of this market which any other industry would be envious of.

At first glance, this evolution is a simple process with no surprises: prices are going to continue to decrease and performance will continue to progress. Customers and the market will benefit from this trend for increasing their equipment by periodically replacing obsolete hardware, whose life cycles are becoming shorter and shorter. The Internet remains the key factor behind this commercial success, whether it be individual consumers who buy PCs so they can access different services and applications which the Internet has to offer, or companies who are revising their investments and changing to the world of web solutions.

The idea of increasing performance and reducing price certainly benefits customers. Suppliers have been obliged to integrate this new and unavoidable constraint in their cheap models: when the price inexorably decreases and when the differentiation between products is only the increase in performance (the idea of the standards of open systems and the roadmap of the silicon industry having excluded all forms of originality in terms of supply), a large number of operators in the computing industry have had to accept a decline in their profits.

Since the end of the 1990s, companies have been rapidly increasing the use of heavy applications with the cost of hardware (servers and workstations) having fallen. Amongst these applications there are: ERP tools, business intelligence tools which are essential in a market that prefers differentiation and direct marketing (such as datawarehousing and Customer Relationship Management (CRM)), or new electronic business applications which are starting to be used in all types of businesses. Add to this the emergence of web structures and the concept of urbanization[3] which is leading to existing applications being grouped into classes of interoperable objects on a very large scale and we can start to understand how the market for servers has experienced, and will experience, a strong demand, even if the volumes are currently being produced to the detriment of the substantial profit margins which designers benefited from during the prosperous years. In other words, not only integrating applications within the same company or organization but also with customers, suppliers and other parties which are part of what is known as meta-organization. Although the period between 2001 and 2004 was marked by a notorious austerity in terms of company investment, there was no slow down in the volume of sales of platforms. The frugality of companies was translated by margin compression as well as by a larger restraint in computing projects and the service prevision that such projects involve.

Requirements in terms of performance, scalability and new characteristics (such as web environment, Java, etc.) have extensively modified the entry-level supply of the server market. At the beginning of the 1990s, the barrier between these categories was quite clear: top of the range servers were specifically designed to serve hundreds, even thousands of users, whereas entry-level servers were built on high-performing PC structures which operated on IBM OS/2, Microsoft Windows NT or Novell Network. Today the simple server-PC-monoprocessor no longer exists. The market has introduced multi-processor machines (commonly eight processors, indeed more) in order to respond to the most demanding performance requirements. When such an abundance of power is not needed, more basic models are offered. However, they generally have the facility of being modular, i.e. scalable, and can follow the evolution of a company's needs (whatever the rate of growth) without any modification of the software. Very specific technology such as service clustering[4] which, up until now, was only available to the top of the range sector of this market has now become accessible to all companies, regardless of their size.

> A cluster is a group of high-speed interconnected machines which work as one unique entity. The fundamental difference between a cluster and a multi-processor system is that the latter functions with just one operating system, whereas clustering technology connects different systems with each of the individual systems running its own operating system. There are three types of clusters (shared

---

3 See the approaches of service-oriented architecture (SOA) in an urbanized system. SOA is an applicable interaction model which uses services in the form of software components. SOA has been popularized with the arrival of web solutions in e-commerce.
4 The challenge to achieve the best performing processors does not only rest with the hardware, but increasingly with the interconnections (inter-node telecommunications) and the necessary software engineering so that these clusters can be high-performing.

> nothing, shared disk and shared everything). Each clustering method functions in relation to how the cluster manages and shares the memory and storage resources within the cluster.

The PC is the dominating technology which allows access to company tools and applications, to email and to other work groups which benefit from the web as well as domestic digital technology (applications and uses, games, audio and video). With this success, the PC is continuing to grow and expand. 25 years after its entry onto the market, the PC has reached maturity and has become a large-scale consumer product characterized by the logic of leveling down in terms of price. In this market, competition now rests on how companies master their production costs which translates into the delocalization of companies to countries where employment costs are relatively low. In other words, it is the changing history of an industry which until recently had strong capital intensity (investments in research and development) to an industry that is characterized by its dominating workforce. After IBM announced that it was going to sell its LENOVO microcomputers to the leading Chinese computing manufacturing company, there is a chance that two of the other 10 largest global PC manufacturers (probably including HP) will stop their production of microcomputers in 2008/09 because of the saturation of the market and the decline in already narrow construction margins in order to concentrate on the more lucrative server market.

The customer, who is a mere observer in this new battle between computer system manufacturers, can only benefit from its advantages. Companies are asking for clear, flexible solutions which are easy to administer so that they can keep up with the economy to ensure the best productivity possible. Whatever the sector, companies big and small wish for ready to use products. Customers no longer want to build their own systems. In addition to this, more and more customers are attracted to the concept of the open operating system such as Linux. The terms and conditions of the future server are clear and the manufacturers are responding; IBM as well as HP, Sun and Dell have translated these terms and conditions into one key objective, which is the ease of use and the capacity in building cluster servers according to needs. In May 2004 IBM presented the eServer Cluster 1350. Ready to use, the system integrates blade servers based on Intel processors joined together in clusters. Presented as easy to install and easy to use, the eServer 1350 functions with Linux. Joining clusters and blade servers allows for a major gain in physical space and in performance. In general, the clusters group together several 100 servers. Due to the reduced size of the blade servers, they occupy less space than rack mount machines.

Since these systems are only applicable to systems that require large processing power, there is only one way in which to invest in the area of supercomputers.

Clustering, which is the base of Linux systems, is no longer confined to research laboratories and will be considered for use by numerous industries. The decline in the cost of this type of machine is one of the main reasons. What used to cost €10 to €20 million 10 years ago represents an investment which in 2006 sometimes did not exceed the €1 million mark.

4.3.2. *Emerging platforms*

These new tools have an increased processing power and storage capacity which are becoming part of the computing world in the 21st century. Today, more often than not, these new tools share the same network infrastructure such as the Internet, local networks and company data storage systems, etc. They offer enormous potential in terms of mobility through devices which are small enough to be held in one's hand with their own energy source, and are capable of communicating wirelessly, having access to everything from absolutely anywhere.

Size, performance (processing power and memory), energy consumption, autonomy and the bandwidth in communication are the challenges that emerging platforms have been able to overcome and continue to improve upon.

The fact that these new platforms are not permanently connected to a network is one of their main advantages. These tools are capable of working either as part of a network or autonomously. Because of this, data must be periodically synchronized between these devices and the network's information or database. Synchronization is a bilateral process of data updating (the data from the mobile station updates the central data and this database then updates the information stored on the mobile station), which is capable of managing competitive access (i.e. when an identical entity is encountered on the two devices during synchronization it is the most recent entity, regardless of whether it comes from the mobile station or from the central base, which will overwrite the oldest one).

Emerging platforms are in fact descendants of everyday electronic devices whose function was initially very specialized, such as video game consoles, decoders for cable television, computer logs for cars, precursors of wireless electronic diaries and mobile telephones. With the increasing ease in accessing the Internet, these specialized tools have little by little integrated a function enabling users to have access to the Internet. The market for devices connected to the Internet other than the traditional PC has experienced one of the strongest growths since 2000.

> Emerging platforms are simple and "closed" systems which integrate hardware, software and connectivity in order to carry out a set of specialized tasks that require access to a network for certain devices (online games, GPS, diary functions). They can be applied to all markets, whether it is about responding to the most diverse needs in the business world or to customer demands.
>
> The only common point that these new types of computers have is in their ability to access the Internet. This simplified approach means that a browser or navigation application does not need to be used and avoids the selection of an Internet Service Provider and the configuration that this involves. Potential users of these new computers who will use them on an everyday basis will no longer need to worry about the selection of an Internet Service Provider since the Internet connections will be easy to establish for the user.
>
> These devices are becoming more and more common in our daily private and working lives, and are progressively replacing the PC which costs more and is more tedious to use for such a simple tool as preconfigured access to the Internet.

The current so-called "emerging platforms" incorporate the most sophisticated methods of accessing the Internet. For less than half the price of a PC, these platforms remain dedicated to specific tasks and their design relies on simplified technical structures. These tools also use simplified operating systems in comparison to those used in PCs and other conventional work stations (for example Windows CE, Palm OS, Sun Java OS, EPOC/Symbian OS to name only the more common ones). In no case will these new platforms aim at competing with PCs in the generalist market. They have the ability to offer a specific service (calendar or address book sharing for a PDA device, video games which enable simultaneous access of several players from over the world) and for this to occur a specific and restricted network is needed. In certain cases, these tools can offer additional services for accessing the Internet (simplified navigation function). Remember that if Internet access from this type of leading device was a competitive advantage, such access would often be developed to define the habits of consumers who, when online, would reveal their intimate habits and routines to the Internet provider. With this information stored in a huge data warehouse, which possesses data mining tools[5], the solutions provider can then organize direct marketing actions (i.e. made to measure marketing depending on the customer), proving to be much more efficient than blind mass marketing.

Emerging platforms use certain types of components and circuits suitable for traditional computers, most often taking into account the requirements of size, price

---

5 Data mining aims at extracting a piece of information from large quantities of data which are stored in a data warehouse. This extraction of information is carried out by automatic or semi-automatic techniques. The source is never directly a data warehouse. An initial stage consists of sending back the raw data (generally obtained on the Internet) to locally store it in a database and then to structure the data.

and the functional specialization of these applications, and these tools are constructed from dedicated ASICs (Application-Specific Integrated Circuits), which are capable of creating more efficient and faster functions for a lower cost than current structures that form the basis of generalist microprocessors.

> An Application-Specific Integrated Circuit is a specialized integrated circuit, i.e. it is designed for a specific task. They are smaller, cheaper, faster and consume less power than a programmable microprocessor. A made to measure graphics chip for a PC can, for example, trace lines or images 10 to 100 times faster than a central processor that is used for general operations. The major disadvantage of ASICs is that they are specialist circuits, a product which is not as readily available as universal microprocessors. Development costs are increased as a result and at least 100,000 units per series must be created each year (except when a made to measure wired solution is required).

Another approach is the system-on-a-chip (SOC) which also allows for very economic and reliable production (like the circuits of quartz bracelets and watches), destined for widely available applications.

4.3.3. *Distributed computing, an alternative to supercomputers*

Current technological requirements in terms of power required by certain applications exceed the resources of only one machine. A solution can be found in parallel structures. Conventional computers execute instructions in sequences. The best performing microprocessors have already introduced the idea of parallelism in the execution of instructions. Without the user knowing, the processor carries out most of the processing simultaneously by respecting the sequential nature of the program. If it is possible to make processors work in parallel, it is possible to extend the concept to a set of microcomputers available on the same network, each of which would carry out different tasks, working on the same issue at the same time. Parallel processing is only effective if the problem to be resolved is naturally compatible with this method, for example, complex vector processing.

The Internet has experienced much more noticeable success than servers, which can be measured by looking at the number of users connected to the Internet. This network uses such an incredible quantity of technical resources which, even today, can be perceived as an abundance of under-exploited resources[6]. Certain companies from the software industry or Internet service companies have already taken advantage of this opportunity in order to freely use the processing power which is

---

6 It is estimated that on average companies only use 40% of their processing power and that home PCs (more difficult to estimate) use barely more than 1/6 of the total power available.

made available by these resources. This is made possible by peer-to-peer (P2P), the next stage in the evolution of the Internet, i.e. the absolute transparence between the sender and the receiver enabling communication without any intermediary or relay of any sort except for the physical nodes of the network itself. The principles of this structure have been used for a very long time in large systems and it is these same principles which govern the functioning of numerous Internet services that we use today without being aware of them. However, some of these services such as Napster and its equivalents, particularly applications which exchange music files online, have contributed to the popularization of P2P (even if specialists in the field consider these P2P methods as a very condensed version of the P2P principle).

> P2P is the sharing of computing capacities on a network which has no intermediary or relay. These computing capacities include the storing and sharing of data between several computers with different processing powers acting as clients and servers in a common space.
>
> A computer which is part of a P2P network also belongs to a workspace and to one or more groups according to the user's needs.
>
> This is specific software which is installed on the client computer, takes care of establishing relationships with its peers on the basis of a common address book and enables a user to look for the resource that is closest to them, or makes data or a certain number of files available to other users.
>
> P2P marks the advent of the user community as a key player in the digital world: a large number of units with very limited resources that work together in a network can become more powerful than large units which have lots of resources.

Distributed computing or grid computing applications also rely on P2P technology. Each client of an application in a grid receives a task to undertake close to a repartition node such as the address or addresses of machines to which the results must be transmitted. Up until now, the principles of computer programming consisted of using exclusively local resources, however, grid technology aims at taking advantage of distant, non-defined resources. The question recalls managing transmission resources as well as computing and storage resources which the company or user that wishes to use them are not aware of. In order to resolve this problem we rely on virtual organization inspired from the layered model of Open Systems Interconnection (OSI) networks.

A first level interface makes it possible to manage processing, storage and network resources. The processing resources work around a mechanism which makes it possible to launch and monitor programs and to control the execution of the results. The storage resources use a control tool of entry and exit files implementing a performance management system. The network resources control the resources allocated to different transfer levels (priorities, reservations, etc.) and determine their characteristics. Versions of the applications are used to manage the uses and updates of the stored resources.

The second level interface ensures connectivity, i.e. the management of communication and authentification protocol, which is required for the specifications of the grid network.

Communication includes the transport, routing and management of names, similar to the Internet protocol (based on TCP/IP, RSVP and on DNS).

The final interface level communicates with the application itself by relying on Application Programming Interfaces (API) and Software Development Kits (SDK) in order to interact with the services and protocols of the grid. The first steps towards a level of standardization have been taken and have enabled the emergence of the Open Grid Services Architecture (OGSA), which is based on Web Services Description Language (WSDL), Simple Object Access Protocol (SOAP), WS-Inspection, and Universal Description, Discovery and Integration (UDDI).

Grid computing, which requires quite a complicated processing method, is currently restricted to specific applications which require intensive computing tools (large universities, CERN, IRM, etc.). However, the results of these pioneering applications have already been very encouraging. The massive development and spread of low-cost Internet access, the progressive decline in equipment and the increasing performance of this equipment makes distributed models today more reliable and more profitable than their competitors, such as centralized supercomputers. The implementation of scientific projects which use models based on supercomputers often face one problem: finance. For example, medium term research and development faces a problem with the constraints of a business case model or economic model whose main criteria is that of return on investment and return on capital employed. Grid computing is an extremely flexible alternative for projects whose long term knock-on effects are unknown and determining a threshold of profitability for these projects is also hard to predict for the future.

Nanocomputers and Swarm Intelligence

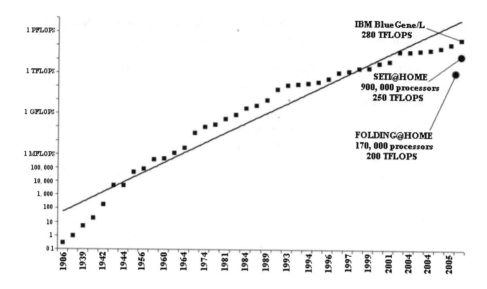

Figure 4.8. *The power of large grid computing compared to the power of supercomputers*

The Decrypthon database is the result of the first comprehensive comparison of all proteins identified in living beings (animals, vegetables, humans), and in particular those with 76 genomes whose sequences are completely understood. This tool will enable researchers to more easily recognize the biological function of the sequences of genes obtained by studies in genomics to establish protein families which have retained a significant level of similarity throughout evolution, as well as the character of a common biological function and to identify regions of these determinant proteins for the creation of their biological function.

Thanks to the contribution of more than 75,000 Internet users and companies who lent their PC processing power for use during Telethon 2001, computations (a comparison of 559,275 protein sequences using the local alignment algorithm by Smith-Waterman) could be carried out in the space of two months. This project, which used the innovative technology of distributed computing on the Internet, was led by a partnership between AFM, IBM and GENOMINING.

Each computer contributed up to 133 hours of processing time, which in total is more than 10 million processing hours. In comparison, the same computations on one standard computer would have taken approximately 417,000 days, nearly 1,170 years. It was the first time in France that Internet users got together on a national level for such a scientific project using the latest information technology, and in particular distributed computing. Prior to this initiative, Berkeley University led the project SETI@home, a distributed computing project using computers connected to the Internet and it was released to the general public in 1999. The objective of SETI@home is to answer the ambitious question: are we alone in the

> universe? This software, which makes it possible to listen to the universe, is in fact made up of two parts: the application itself and a screen saver. Via the Internet, the application downloads blocks of data which come from telescopes from all over the world and then the application subjects this data to mathematical processes with the aim of marking the presence of synthetic radio signals from other worlds in the universe, worlds with intelligent, technical sufficiently developed beings. The results are then transmitted and the application acquires new blocks of data.
>
> SETI@home is today the most powerful application of distributed computing in the world with an estimated processing power of more than 250 TFLOPS which is supplied by more than 900,000 machines in comparison with BlueGene/L's 280 TFLOPS[7], which is one of the largest supercomputers of the moment.

4.4. Advanced platforms – the exploration of new industries

In the previous chapter we introduced several research methods which one day could lead to the replacement of conventional transistors by equivalents built on a molecular scale (see section 3.6). There would be no reason to question the processing techniques of a computer that would use such components instead of traditional silicon circuits. We would still use programs made up of arithmetic and logic instructions, conditional loops (if...then...), storage directives and information restoration. These same instructions make it possible to translate software programs into binary instructions which are interpreted by the machine. In other words, hardware could be dramatically reformed using hybrid or new technology, but the principles of software production that these machines use would remain the same. We will continue to resolve a problem and to respond to a request by relying on the same approaches in terms of analysis and programming.

The main advantage of this type of program is the independence of the logical program in relation to physical platforms and processors. Today algorithms are universal and in the majority of cases, source programs or even executable codes (for example Java structures) can be exploited on almost any structure available on the market. This universality has become one of the cornerstones of modern computing. The evolution of current information systems, the interoperability of programs and executable codes, in any type of environment as we have just seen, is becoming more and more diverse.

---

[7] IBM Blue Gene/L performs at a peak of 367 TFLOPS and has a sustained steady performance of 280.6 TFLOPS.

> An algorithm is a procedure of resolving a problem set out as a series of operations to be carried out. The implementation of an algorithm consists of transcribing the algorithm into operations which are similar to a programming language (also called source code) in order to be interpreted or compiled i.e. translated into a language understood by the machine (object code).
>
> The science of algorithms or Algorithmics is attributed to the Persian mathematician Al Kwarizmi whose name is behind the origin of the word. In 1637 René Descartes put forward an interesting approach relating to this discipline in his text the *Discourse on the Method* in which he stated: "To divide each of the difficulties that I was examining into as many parts as possible and as might be necessary for its adequate solution".

In more specific cases, new computation and processing devices could be part of a decisive and radical alternative to Boolean computers. Based on technology which is different to that of semi-conductors, these new systems today remain in the laboratory stage of development. They are still far from being used industrially as their practical applications are still uncertain. The majority of these solutions will more than likely not be developed at the industrial level in the next 10-15 years (if the industrial quantum computer one day exists it will probably not be within the next 20 years). However, to overlook the potential of these devices would be the equivalent of having ignored the industrial future of the airplane in the 1890s when Clément Ader's Flying Machine "Eole" struggled to take off.

These non-Boolean information systems (i.e. those operating with a processing logic different to that of binary computing) include the DNA computer, the optical computer[8] (which uses photons with no inertia instead of electrons) and include applications such as the quantum computer which are the result of fundamental research. This solution offers unprecedented processing potential by using traditional methods. However, they are extremely limited in their usage. In fact, their use relies on very specific algorithms, and these algorithms are limited in number and confined to the resolution of extremely specific problems. The examples of the computers above have been mentioned in particular because they will form part of computers over the next 20 years and will be essential for use in specialized functions such as cryptography and optimization under constraint. Except for particular cases, the computer of the next two decades will more than likely remain Boolean regarding its general functions. However, there is a large chance that computers will soon possess new resolution methods in data processing.

---

8 The optical computer is a parallel computer which is useful for specific applications, for example, in Fourier transforms.

4.4.1. *Quantum information systems*

"He who is not shocked by quantum theory has not understood what it is", said the physicist Niels Bohr in order to make people aware of how their attitudes concerning the laws of physics undermine the infinitesimal.

At the moment, is it possible to imagine that in a quantum world the same particle can exist in two places at the same time? How can an electron pass through impenetrable walls with the help of other particles without scientists really understanding why this occurs? How can this electron find itself in different states of matter thus making the dichotomic logic which constitutes the founding principle of our current processors almost impossible?

Rather than considering the fact that the strange behavior of electrons on a nanometric scale poses a problem, researchers and engineers find this an advantage towards developing a future generation of computers with an increasing number of applications.

Without discussing in detail the main principles of quantum physics, which govern the functioning of these new devices, it is useful to mention some essential points.

In traditional physics, it is possible to simultaneously know the speed and the position of a particle or object (i.e. a planet, an airplane, an ant, etc.). However, in quantum mechanics it is impossible to simultaneously indicate the speed and position of an electron because of the uncertainty that the observation introduces. This is the idea of Heisenberg's uncertainty principle which states that it is impossible for a particle to have a well defined momentum and position. It is only possible to predict states of electrons.

These states can be superimposed and these superimpositions can be integrated together (in other words, quantum intrication). Using such a principle in a system would offer processing possibilities whose power would exponentially increase in relation to the alternatives of conventional binary electronics.

It is theoretically possible to assemble up to 40 basic units of qubits. Beyond this limit it is believed that the quantum system bonds with its environment and that its properties are no longer exploitable. This enables such a system to process in parallel the equivalent of up to 2^{40} conventional bits. This is the absolute maximum in terms of available processing power which goes well beyond what is currently possible.

> A qubit (the contraction of quantum and bit) is a two-level quantum system which is created in the laboratory. The qubit is the quantum state which represents the smallest quantum unit of information storage. It is made up of a superposition of two spin-based states, conventionally denoted |0> and |1> (pronounced ket 0 and ket 1). A qubit state is made up of a linear quantum superposition of these two states. It is manipulated by applying an alternating electromagnetic signal which can be imagined as a fictional field acting on the spin.
>
> The concept of a qubit memory is fundamentally different from a conventional memory: a bit can only have the value 0 or 1 at any given moment. The qubit can have both values at the same time. However, this combination is not a simple mix resulting from probability as is often thought; the coherence of the qubit must be insisted upon because it is this coherence which guarantees that the quantum information remains intact. A qubit must therefore be uncoupled from its environment in order to maintain its coherence and thus the integrity of the information.
>
> A qubit is only used to provide a result when a state is degenerated as pure 0 or pure 1, without which it only provides a random bit. Its state is destroyed by reading (remember that the state of an ordinary bit is only destroyed by writing). Reading a qubit which is initially in one state will change its value.

In sensitive areas such as cryptography (founded on the decomposition of prime factors) or in certain vector applications which require very large parallel computations, the limits of conventional computers are starting to appear in terms of response time. Even when applying Moore's Law and the roadmap of the silicon industry whereby hardware is going to increase its performance throughout the next decade, the resolution of certain problems would lead to processing times of several hundred years with even the most advanced traditional computers available. For approximately 20 years, scientists and researchers have wanted to use the properties of matter on a quantum level. The first person to promote this concept was Richard Feynman who suggested its use in order to resolve problems referred to as NP-hard (i.e. where the number of calculations necessary to resolve one of these problems increases exponentially with the number of variables concerned, which leads to an insolvable problem on a traditional computer).

Quantum computing is part of a new research area found at the crossroads of the theory of traditional computing, quantum computing and quantum mechanics. The discovery of extremely powerful quantum algorithms has more recently led to much interest in the area of quantum computing. The rapidity of certain operations, such as the factorization of large numbers with quantum computing, can be exponentially increased in comparison to traditional binary computing.

Whether it be traditional or quantum, a computer always manipulates binary information coming from physical variables which are capable of having two distinct states, represented by the values 0 and 1. However, the difference between a bit and a qubit is that at any given moment a bit can either be in state 0 or state 1 whereas the qubit can be simultaneously in both states. For example, the qubit can be worth 0 with a probability of 17% and worth 1 with a probability of 83%.

This concept affects one of the big mysteries in the world of the infinitesimal: an atom can simultaneously spin in both directions, an electron can be present in two different energy levels and a photon can be polarized in two directions. In the same way, while a traditional two bit binary register can only take one of the four distinct states (00), (01), (10) or (11), a two qubit register can superimpose these four states coherently at the same time, each state being associated with a certain probability. A three qubit system is therefore a coherent superposition of eight states (2^3), and a forty qubit system has a coherent superposition of 2^{40} states. In order to illustrate the magnitude of these figures, a 270 qubit system would possess more configurations than there are atoms in the known universe and a modification to only one of these qubits would affect the 2^{270} states of the system at the same time. This parallelism has no equivalent in the traditional world.

In other words, since the number written in a register can have several values at the same time, the manipulation of such a register in a quantum computer makes it possible to simultaneously explore situations corresponding to the different values of the register. The information which is extracted from the calculation takes advantage of the effects of interference. The obtained result will then depend on the different paths taken by the different paths of the register. The quantum superpositions can be inseparable, i.e. only the state of the whole register is known without determining the state of one qubit. The quantity of information contained in these inseparable states is exponentially larger than in a traditional system of the same size. This quantum parallelism enables the exploration of a much larger space for a given number of operations.

> The Schrödinger's cat paradox (1935) is a pure thought experiment which illustrates the absurd character of quantum physics the moment scientists try to apply the laws of quantum physics to macroscopic objects. It was designed in order to highlight the problem of measurement: it is the measurement or observation and the fact of branching off from one superimposed quantum state to an observed or measured quantum state which perturbs the system.
>
> A cat is enclosed in a box with a flask of deadly poison. In a corner of the box there is an atom of radioactive uranium as well as a detector designed to function for only one minute. During this minute there is a 50% chance that the uranium atom will disintegrate by ejecting an electron which will set off a mechanism emptying the contents of the flask. The question is if the box is closed and the experiment is launched will the cat be dead or alive?
>
> The law of probabilities states that there is a 50% chance that the cat will be dead

> and a 50% chance that it will be alive. According to the laws of quantum mechanics the cat will be dead and alive at the same time (the cat would be in a superposition state in which it would accumulate several states which are macroscopically incompatible). According to these same laws, the uranium can be both complete and disintegrated at the same time, implying that the atomic particles can exist in several superimposed and simultaneous states, similar to the idea of parallel worlds.
>
> This paradox is used in what is known as decoherence, a principle which affirms that the superposition state can only be maintained in the absence of interactions with the environment that sets off the choice between the two states (dead or alive). In quantum physics, when there are two possible states for a particle the combination of both these states is also a possible state. However, the observation itself will lead to a reduction of just one state. If it is possible to succeed in creating a direct dependence between the state of the particle and the life of the cat, the cat could be placed in a superimposed state where it would be both dead and alive until observation which would reduce the cat to just one state, i.e. dead or alive.
>
> The information contained in the register of a quantum computer exists simultaneously at state 0 and state 1 superimposing the two states at the same time, like Schrödinger's cat. If it is possible to isolate such a system from interacting with its environment then it would be possible to simultaneously maintain this register in the two states.

Quantum mechanics makes it possible to carry out processing tasks which are impossible for current computers, even after taking into account the future progress in their processing power. Quantum algorithms exist which resolve problems in polynomial time, traditional algorithms resolve the same problems in exponential time. In 1994 Peter Shor from the AT&T laboratories designed an algorithm making the most of this property for factorizing very large numbers in polynomial time which signifies, in mathematical terms, that the growth in the size of encryption keys would no longer be an unconquerable obstacle.

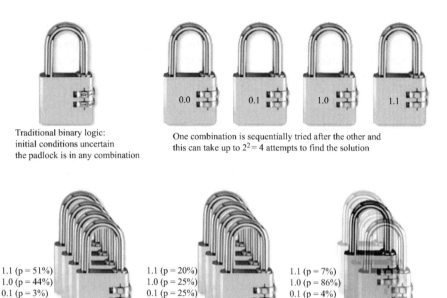

Figure 4.9. *In 1996, Grover came up with a quantum algorithm which finds a solution in a table of N elements in O (square root of N) attempts. Grover's algorithm improves the processing time of all easily verifiable problems. In conventional logic, when an object N is being searched for, it takes on average N/2 attempts to succeed. In quantum logic, by relying on Grover's algorithm, it takes approximately N^2 attempts. For example, in the case of a quantum two bit padlock, Grover's algorithm will find the combination during the first attempt (N/2=1)*

The experimental verification of the problem of the quantum two bit padlock (shown above) was undertaken in 1998 by Isaac Chuang[9]. He built the first quantum chloroform-based two qubit computer, which is capable of searching the different periodicities of a function. With this machine, Chuang also implemented Lov K. Grover's algorithm enabling him to find a piece of specific data amongst four pieces of information in only one phase. The following year he created a three qubit machine (based on eight elements).

---

9 It should be pointed out that Chuang's qubits are based on nuclear magnetic resonance. Such a system is not scalable and nuclear magnetic resonance quantum systems along with many qubits will never be created.

In 2000 he created a five qubit computer by using five fluorine atoms of a specially designed complex molecule. With this system it became possible to find the correct combination between two elements in only one phase whereas with a traditional approach it took up to four phases to find the correct combination (two of the five qubits are used in the research of the solution, the three others are used in the computation of the result). Three years later, the same team from IBM's Almaden research center succeeded in the factorization of a number with the help of a seven qubit quantum computer by carrying out the experimental demonstration of Shor's algorithm. This experiment used nuclear magnetic resonance applied to seven spin synthesis molecules (i.e. a nucleus with five fluorine atoms and two carbon atoms) in a liquid environment. These molecules can interact with other molecules like qubits and are programmed by radio wave pulsations with the help of nuclear magnetic resonance.

> Shor's algorithm is a quantum algorithm which is used to factorize a number N in time $O((\log N)3)$ and in space $O(\log N)$, i.e. it is used to resolve problems in polynomial time for which traditional algorithms require exponential time.
>
> Many public key encryption systems, such as the RSA algorithm for public-key cryptography, (largely used in e-commerce and even more so for the secure exchange of information on the Internet) would become obsolete if Shor's algorithm was one day implemented into a functioning quantum computer. With the RSA algorithm, a coded message can be decoded by the factorization of its public key N, which is the product of two prime numbers. It is known that traditional algorithms cannot do this in a time $O((\log N)k)$ for any value of k as they very quickly become unworkable when the value of N is increased. Shor's algorithm has since been extended for use in other public key encryption systems.
>
> Like all algorithms for quantum computers, Shor's algorithm is based on probability: it gives the correct response with a high probability level and the probability of failure can be reduced by repeating the algorithm.
>
> Shor's algorithm was demonstrated by a group from IBM who factorized 15 into 3 and 5 by using a seven qubit quantum computer.

While conventional transistors make up switches which enable the development and creation of traditional bits, the manipulation of qubits requires new switching devices to control this quantum information, which is created by the orientation of the magnetic field of simple atoms.

Experimental approaches firstly aim to create a basic logic gate, i.e. coupling two quantum systems so that the state of one modifies the state of the other, and secondly to progressively increase the number of these coupled systems in order to form a genuine computer.

It is easy to imagine the difficulties relating to the confinement in a volume of a dozen (or more) qubits and, furthermore, the difficulties relating to the surrounding temperature (by taking into account the difficulties faced when manipulating such systems, the maximum number of qubits which can be implemented into the most recent prototypes of quantum computers is no greater than seven because the magnetic signals which measure the orientation of the spins and determine its value: 0, 1 or both become weaker as the number of qubits increases).

Here we are dealing with the most delicate problem that is hindering the development of the quantum computer: the difficulty of maintaining more than one particle at a time in the superposition state. The localization or impulse of a quantum particle in the superposition state can only be defined by statistical probability which ensues from the particle's wave function.

In order to exactly know what these values are, the particle must be treated with an instrument which has large number of atoms. However, the wave function collapses and the observer is only left with one of the two values, the other value is definitively used by applying the uncertainty principle. This is also known as the phenomenon of quantum decoherence[10].

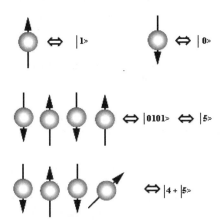

Figure 4.10. *Coding numbers with qubits. Qubits are represented by magnetic spins. An upward spin represents state 1 and a downward spin represents state 0. With a four qubit register the figure five can also be coded for example. If the lightest qubit is in the superposition state, its spin will be horizontal and the register will represent the figures four and five. The majority of other superposition states require the intrication of several qubits which cannot be displayed (source CEA)*

---

10 The main result in this domain is that it is possible to compensate decoherence if it is quite weak. It has been shown that an error rate of 1/10,000 per logic gate makes it possible to use quantum error correction codes; however, this is at the price of a high level of redundancy.

Thus, one or several qubits retain their quantum character and are able to work in the superposition state, and they must be isolated from all matter and energy that they interfere with.

Today, however, by using different techniques some teams have said that they have been able to maintain the quantum state for short sequences of bits (four to seven at most) and for durations of time that are sufficient enough to enable the creation of a few operations (some tenths of a microsecond).

Scientists must remain realistic but continue to persevere with their ambitions. The common applications coming from this research will only be available in the long run (if they emerge at all) i.e. more long term than the advent of molecular informatics, even if some very specific computing applications stemming from quantum theory emerge in the nearer future.

How can the quantum computer be programmed?

First of all, a calculation must be developed which will lead to a non-superimposed state. If a qubit which is found in a superposition of states is searched for, the response will be random and will not lead to anything interesting. An algorithm leading to a unique response for all paths of possible calculations must be developed.

It is a problem which is similar to that of enigmas where a true response must be obtained by setting a series of intermediaries, some of which we know always lie and others never lie.

The question is therefore to find a calculation which achieves this invariant, for example, in the case of breaking the code of a ciphering key that we are looking to crack.

The second problem consists of measuring the quantum state: the reading of only one bit in a quantum state destroys the complete state. The number of times that the calculation must be repeated is equivalent to the number of bits that the requested response possesses, e.g. if the requested response has two bits then the calculation must be repeated twice. However, the number of times the calculation must be repeated is only proportional to the number of bits and is not exponentially larger, which is the exact objective that is being researched.

In what shape might the quantum computer, whose promoters predict that its first industrial applications will appear around 2020, be? Will its processing unit be solid or liquid? Will it be a large system like the very first computers were or will it be a pocket-size device?

If the first experiments benefited the use of magnetic resonance applied to synthesis molecules in a liquid environment, other approaches also exist, for example, that of ion traps which also require a liquid environment.

However, the development of an operational machine has encouraged laboratories to focus on methods exploiting qubits fixed on solid substances or on prisoner photons in optical cavities (the spins of electrons confined in semi-conducting nanostructures or the spins of nuclei associated with mono-atomic impurities in a semi-conductor; see section 3.6.5).

The CEA Quantronics group succeeded in producing a solid state quantum bit. Quantronium uses three Josephson junctions. A Josephson junction is made up of two superconducting electrodes separated by a very thin insulating barrier in order to allow the passage of pairs of electrons by the tunneling effect.

Two of the junctions form the quantum bit, the third is used either as an insulator or as a measuring device when a current crosses the junction. By sending radiofrequency impulses to the chip, researchers have succeeded in writing the superimposed states of a qubit. They then read them by applying current impulses to the reading junction and by measuring the tension at their terminals (allowing them to find out that the qubit will have lasted 0.5 microseconds).

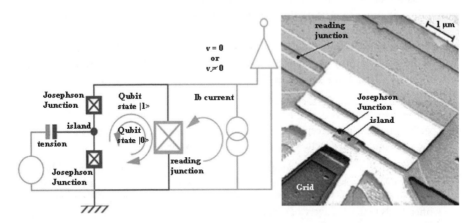

Figure 4.11. *Quantronium: two small Josephson junctions which establish a small island and a reading junction are placed in a loop. The control parameters for quantronium are the tension applied to the grid electrode combined to the island, and the magnetic flow throughout the loop. For the qubit, the reading current Ib is equivalent to a supplementary flow which develops the loop current of the two qubit states: $|0>$ and $|1>$ (source CEA, Quantronics group)*

Several groups have developed solid-state qubits, be they either superconductors (at the CEA, in Grenoble, at Delft, in Gothenburg, at Yale University, at the National Institute of Standards and Technology (NIST), in Berkeley University and in Santa Barbara), or semi-conductors (at Delft and at Harvard University).

Quantronium is the first qubit which has enabled the reproduction of fundamental experiments of quantum physics.

The next phase consists of depositing two of these qubits in an operating device such as a logic gate. Such a logic gate has been used by John Martinis' team from the University of Santa Barbara (UCSB).

However, extreme caution is needed since this is exploratory research. The achievement of solid-state quantum computers is far from being reached.

4.4.2. *DNA computing*

Nothing limits the design and production of computers apart from electronic technology. Up until now, all concepts explored to continue the process of miniaturization have been founded on the same principle used by the computers of today, using electrons as carriers of information, as well as transistors or their derivatives in order to process this basic information.

Yet on a nanometric scale, a computer can function in a completely different way. One of the most iconoclastic ways considers using the carriers of genetic information of living organisms, i.e. DNA.

> For millions of years nature has been resolving all the problems of life on Earth, such as movement, perception, protection, camouflage, etc., by continuing to use the same processes: genetic variation and selection variation.
>
> The search for a solution through the help of a genetic algorithm is incredibly remarkable because the evolving algorithm is practically blind. Starting with nothing but the knowledge of how to compare different solutions, the algorithm often leads to the discovery of excellent solutions.
>
> Genetic algorithms question our idea of intelligence because we are not used to thinking in terms of evolution.

The molecules of life are capable of storing unbelievable quantities of data in sequences made up of four bases called adenine, cytosine, guanine and thymine (represented as A, C, G, and T respectively) and enzymes can manipulate this information in a high level of parallelism.

The potential of this approach for a new generation of processors was highlighted by Leonard M. Adleman in 1994. He demonstrated that a DNA computer could, in fact, resolve problems which were unsolvable for conventional computers, such as the sensitive problem of the sales representative or a variant of this, the problem of Hamiltonian's Path (the name comes from William Rowan Hamilton, a royal astronomer in Ireland in the 19th century) and which are sufficiently simple for a prototype of a DNA computer to resolve. However, these findings could not form any concrete proof of the performance and of the potential that the processor industry perceives.

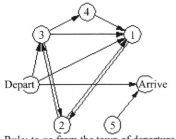

1. Rule: to go from the town of departure to the town of arrival by only going through each town once and by using the available routes

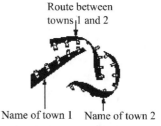

2. According to the technique of molecular biology, DNA sequences for the towns and for the routes that link them are identified. The route sequences are complementary to the town sequences in their extremities

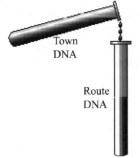

3. By a process of enzyme amplification (Polymerase Chain Reaction) the sequences of DNA are copied millions of times and then are mixed together in a test tube where they combine because of their complementarity

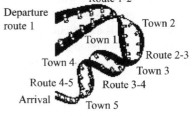

4. When the combination has been carried out the process of filtering is next. Filtering aims to extract the sequence which will lead to the correct solution (i.e. the strand departure route 1, route 1-2...route 5, etc.

Figure 4.12. *The resolution of the sales representative problem by DNA computer*

Science is far from being able to create molecular machines from nothing which are as perfect as DNA polymerase. However, scientists can use cells which are available to them and which have been generated by more than three billion years of evolution. Modern genetic engineering already uses certain types of these molecules. Even if evolution has never produced cells which play chess, scientists can imagine building machines with interesting processing capacities in order to resolve the most diverse problems by using the tools that nature has made available:

– base pairing: the sequence of a chain represents a piece of information which can only be associated with the complementary sequence. It is this property which is the foundation of DNA computing. Bringing a billion billion ($\sim 10^{18}$) molecules into contact inside a test tube in order to get them to pair recalls the idea of carrying out large-scale parallel processing. When a strand of DNA comes into contact with complementary DNA, the two strands wind around each other in order to form a double helix. The strands do not chemically react but bond because weak forces (hydrogen bonds) are established between the complementary bases. When a strand of DNA comes into contact with another strand of DNA which is not complementary or which does not possess long complementary parts, then the two strands do not bond. All these possibilities of association are explored automatically and naturally;

– polymerases: polymerases duplicate the information of a molecule by creating a complementary molecule. For example, DNA polymerase produces a complementary strand of DNA from any molecule of DNA. The DNA polymerase begins the copy process when it comes into contact with a starting signal which indicates the beginning of the sequence to be copied. This signal is supplied by a start codon, which is a particular, often short sequence of DNA. As soon as the DNA polymerase finds a codon paired with the model, it begins to add bases from the codon in order to create a complimentary copy of the model;

– DNA synthesis: the technique of polymerase chain reaction (PCR) makes it possible to multiply a sequence of DNA millions of times in a few hours, to receive a sufficient quantity of them in order to proceed to sample analysis and to highlight the researched results.

In order to analyze the DNA samples containing the result or the solution of the problem, the principle of electrophoresis is used. This process is carried out in agar gel cut into wells where the DNA solutions which are to be analyzed will be placed. The buffer is slightly basic with a pH of 8.6, and the DNA ionizes itself on its phosphate groups and migrates towards the anode. The resistance of the porous gel to the movement of the DNA in the electrical field is proportional to the size of the DNA fragments. They are therefore separated in relation to their size expressed in base pairs, the smallest fragments that move the fastest. In order to have an idea of size, the following are placed inside one of the wells: electrophoresis gel, a marker, and a mixture of fragments (the size of which is known) which is generally obtained by the digestion of DNA that can be perfectly identified by a restriction enzyme.

A DNA-based computer is primarily an assembly of specially selected DNA strands whose combinations will result in the solution of a certain problem. The technology is currently available to choose the initial strands and to filter the final solution. The major force of DNA-based computing is its enormous capacity in parallel computing: with a certain organization and sufficient DNA it is potentially possible to resolve complex problems that are sometimes reputed as being unsolvable by parallel processing. It is a much faster process than the traditional process for which large level parallelism requires an increase in the number of processors to be used, which is expensive. In the case of DNA computing, in order to increase the level of parallelism it is only necessary to increase the amount of DNA. It is therefore a simple question of amplification. Several applications of this research, especially in the medical or cryptographic fields, are already being studied.

One of the promising methods which several teams have been working on is the development and self-assembly of the bricks of DNA in order to construct electronic nanocircuits.

> How can a DNA-based computer be capable of resolving complex problems such as factorizing numbers, directing messages in a telephone communication switch or evaluating a financial risk?
>
> In the 1930s, mathematicians demonstrated that only two components were enough to construct a computer: a means for storing information and a tool for carrying out simple operations. The legendary Turing machine[11] prefigured current computers which store information in their memory in the form of a succession of binary values and manipulate the information with the help of their arithmetical logic unit. No other means capable of storing information and possessing a processing unit with all its capabilities can be used like a computer.
>
> A processing system using DNA relies on fundamentally different coding mechanisms to those of a conventional computer: in traditional processing systems it is the information system which leads to the information being coded in binary form. With DNA-based computers, the information is translated in terms of the chemical units of DNA.

---

11 The Turing machine would inspire the beginning of computing by developing a device capable of resolving every problem which can be described in algorithmic form. This abstract machine, reduced to a ribbon (the memory which stores information) and a pointer which moves on the ribbon that can read and write symbols. By following instructions from the controller through to the finished states (a table of actions which indicates to the machine what symbol to write, how to move the reading head, left or right, what the new state is, etc., and a register which memorizes the current state of a machine), the machine will resolve the set problem by following an algorithm.

> The principle of processing with a DNA-based computer consists of synthesizing particular sequences of DNA and letting them react inside a test tube. The logic connective AND can be obtained by separating the strands of DNA according to their sequences, and the logical connective OR can be created by depositing together the DNA solutions which contain specific sequences.
>
> In order to resolve problems such as the famous Hamiltonian Path or sales representative, Adleman first of all produced a DNA solution in which the DNA molecules were conventionally coded, each path of the DNA molecules was coded between two points. By a process which alternated the phases of separation and amplification, the scientist eliminated the impossible paths, such as those which contained points that were not supposed to contain or even insulate the correct solution.

By trying not to ignore the unquestionable advantages of DNA-based computing in the resolution of insolvable problems with traditional processes, it should be pointed out that numerous obstacles still remain. The following problems should be mentioned: base pair mismatching in the process of assembling pairs, and the excessive quantity of required nanoelements in order to carry out the simplest of calculations (even if these elements are easily accessible as mentioned earlier in the chapter). In fact, it is believed today that DNA computing will more than likely join other methods of nanoelectronics where it will occupy a very specific and limited role in our vision of the information systems of the future.

Chapter 5

Elements of Technology for Information Systems of the New Century

5.1. Beyond processors

Restricting future trends and the next generation's computers by focusing only on the processor would be over simplistic. Apart from the race towards higher performance and the miniaturization of circuits, the progress of information systems takes place principally in the production of computer memory and mass storage. However, even this "turgescent" vision does not sufficiently characterize the development of computers over the next two decades.

Since the computer has become portable and a tool for communication, it has been integrating itself into our world in an ever more natural and comprehensive way. The schematic or, in other words, simplified structure of a computer can currently be divided into the following five blocks: the processor, memory and mass storage, telecommunication networks which allow the interaction with other machines, power supplies which now predominantly allow mobility and web diffusion and, last but not least, the interface between the user and the machine or the machine and its environment.

Imagining the computer of the future requires imagining a joint development of these five elements of technology as they often have a strong impact on each other.

136   Nanocomputers and Swarm Intelligence

As previously done for processors, all technologies need to be sub-divided into their current industrial applications and their long term perspectives out of which some may always remain in the area of exploratory research. The combination of different available solutions leads to a chronological outlook on the development of computers and information systems over the next 15 years.

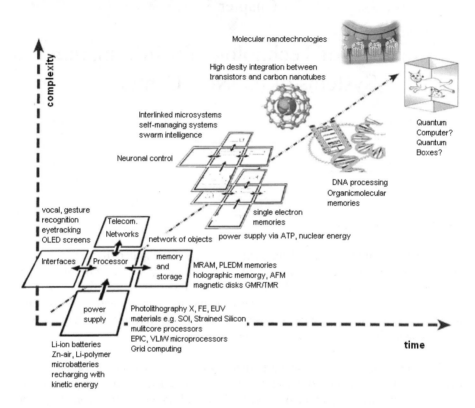

Figure 5.1. *The roadmap of information systems over the next two decades*

The roadmap of the information systems for the next two decades is based on five dimensions:

– smaller, faster and more powerful processors which use traditional silicon technology but remain open to new developments. These new developments may find complex applications by leading companies into the fields of payment mandates, cryptography, expert systems, etc. Current applications will then become technologies of the past;

– memory which mainly aims for the smallest possible number of electrons needed to store one bit. Currently, several hundred thousand electrons are "discharged" in the process of storing one single bit. The ultimate goal is the development of memory that consists of only one electron and is also non-volatile, in other words, it does not need a periodic refresh rate to maintain the information saved. New technologies will therefore progressively reduce the number of electrons used, increase the speed of transferring data and decrease the electricity consumption. In the field of mass storage, the density of the information stored will play a predominant role as today all information is stored on digital devices. Research is mainly concentrated on three-dimensional storage which saves information as an entire "page" during each transfer. It should be kept in mind that current storage devices are two-dimensional, which means that the information is cut into small chunks and applied to the surface of a magnetic or optical disk, as was the case for magnetic tapes and punched cards. These new solutions take their inspiration from the living world: information is stored in the three dimensions of a protein. Similarly, in the field of optics, data is stored in a three-dimensional hologram in the center of an extremely small crystal;

– the third axis of progress is that of power supply, which has become crucial since systems have become more and more portable. The race towards the independence of power supplies and the importance of weight began with the appearance of laptops, cell phones, PDAs and other digital portable devices. However, with the notion of web diffusion the problem has increased. Currently, "intelligent" devices are extremely small and often permanently integrated into their environment (e.g. biochips and microsystems), so that replacing a power supply would be very complex and therefore virtually impossible. These new power supplies will perpetuate progress in the field of electrochemical batteries (e.g. lithium polymer batteries which are extremely slim, zinc-air batteries in which particles of zinc are mixed with an electrolyte). Furthermore, they allow for a density of energy which is 100 times higher than that of traditional batteries. In other words, their lifespan is just as long as the device's lifespan. These power supplies cannot exceed the dimension of the sub-millimeter and need to be rechargeable while working, for example, the on-going production of energy through mechanical vibrations;

– the fourth axis of development concerns the interfaces with the user or the environment interfaces. From a simple keyboard and screen, the computer has developed in around 40 years into something that understands all five human senses and nearly all natural movements. The application of speech recognition on a large scale is now just as common as handwriting recognition. The impressive progress in these fields will allow for smoother, faster and more natural exchanges with machines. Information will be always shown on some form of screen. Current screens using LCD or plasma technology, as new as they may be, will soon be overtaken by less energy consuming technologies. Screens will also be even flatter.

The use of a backlight will end and be replaced by ambient light with a better use of the light in the surrounding area. These screens can modify their shape and can be folded just like a piece of paper. In the long term, there is also the possibility of creating an interface that senses and gestures. This interface which directly connects the brain and the machine introduces the notion of telepathic control. For a long time these applications were thought of as a device which would be implanted into our body and would therefore require an operation. Their application was perceived as ethically unacceptable. However, new and very promising applications which will not be implanted into our body have been developed and could enter the market in about 5 to 10 years' time;

– the fifth axis of development is that of the communication between systems, in other words, networks and telecommunications. If the Internet mainly consists of the global interlinking of traditional networks, the next step, which will favor the development of ambient intelligence and web diffusion, is being developed on the level of large-scale wireless networks. The specialty of these new networks lies in the fact they no longer require permanently installed antennae. These networks consist of active nodes or, in other words, antennae which are directly integrated in the terminals (i.e. the objects that interact with each other). These telecommunication systems will be able to manage themselves. This analog principle known as P2P (see section 4.3.3) is perfectly adapted to the introduction of a highly interlinked network of objects that constantly interact with each other.

5.2. Memory and information storage systems

In the development of digital appliances, one very important factor is the incredible increase in the memory required to execute commands as well as to save data on a permanent basis. Multimedia files which are becoming increasingly large in terms of memory (sound, photos, videos, etc.) are largely responsible for the constant need to increase computer storage capacity. Future generations of machines will accommodate this tendency in the fields of RAM and mass storage.

The technology of magnetic data storage will have followed the same development as the semi-conductor. The speed of computers and the density of data storage has increased one hundredfold between 1985 and 1995. The same factor was also observed between 1995 and 2005. However, in the field of silicon we are about to come up against a "memory wall" in the development of magnetic storage of data. In other words, in the short term current technology will have brought the density of information storage to its maximum capacity. This limit is super-paramagnetic, which means that the binary data stored are positioned so close to one another that their magnetic polarities start interfering with each other. An even denser storage of data would mean that non-magnetic technology would need to be applied. The research, whose results are most likely to enter the market in the near future, relies on holographic memories or the usage of devices which derive from the use of atomic force microscopes (AFM) for mass storage.

Furthermore, memory in current computers relies on the globally applied technology of semi-conductors which was mentioned above. It is subject to the same physical limits as processors and might also be replaced by other solutions. Increasing the performance of computer memory requires a reduction in the number of electrons needed to store a single bit of information. From the very beginning of information technology the objective of minimizing the electrons used for the procedure of storing on a single bit has been the decisive factor in the development of storage devices. Today, several thousand million electrons are required to store one piece of binary information. The ultimate goal would be that only a single electron is used. Other approaches simply spare themselves the trouble of working with electrons. They exploit the properties of photosensitive materials such as three-dimensional organic memories (see section 3.6.3).

In 1997, the largest database was Knight Ridder's DIALOG with a capacity of seven terabytes (according to www.SearchDatabase.com).

In 2002 the largest database was that of the Stanford Linear Accelerator Center which allowed for the storage of more than 500 terabytes of experimental data, which was close to 5% of that year's global storage capacity.

Techniques of parallel processing as well as optical interconnections help to resolve the limited speed of data transfer in computers that could easily adopt hybrid technologies.

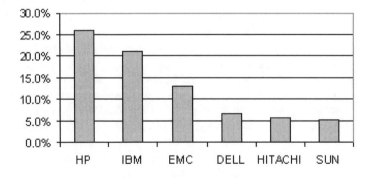

Figure 5.2. *The market of mass storage (source IDC – 2003)*

5.2.1. *Memories consisting of semi-conductors – perspectives*

RAM or today more precisely DRAM (*Dynamic RAM*) is a space that allows the temporary storage of data while the program is running, as opposed to storing it on a mass storage memory such as the hard disk. DRAM needs to be "refreshed" periodically to preserve the stored data. This is why it is called *dynamic*. Once the computer is switched off all data saved in this memory is erased.

> A DRAM memory unit consists of a MOS transistor and a capacitor. Once the transistor is activated it becomes permeable and charges the capacitor. DRAM cells are situated on a lattice and at the intersections of lines and columns. It is therefore possible to either "read" the tension of the capacitor or "write" the logic value by creating a difference in the potential of the capacitor's terminals (e.g. +5 V or 0 V according to the value to be written).
>
> The capacitor discharges at some point because of leaks in the electric current. This is why the work cycle of the memory includes a refresh period of the data which consists of reading and simultaneously rewriting the logic state of all points to be saved in one row.

SDRAM (*Synchronous Dynamic Random Access Memory*) is the next generation to follow DRAM. This type of memory synchronizes the entry and exit of signals at the rate of the control bus of the processor in order to optimize efficiency. SDRAM, which appeared in 1996, is now hardly used. It has been replaced by the DDR RAM (*Double Data Rate RAM*) which is synchronized twice during the work cycle, rising and falling at the front during each clocked work cycle. DDR RAM is therefore twice as fast as the SDRAM.

DDR2 memory

core frequency = 100 MHz   work cycle = 200 MHz   data frequency = 400 MHz

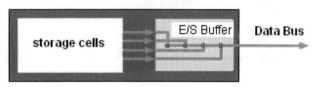

Figure 5.3. *DDR2 is the next step in development and doubles the performance of DDR RAM. Being faster, it also offers a higher capacity because if the current DDR circuits are limited to 1 Gbit, DDR2 allows for the capacity of 2 Gbits or 4 Gbits. With the increased frequency, memory cells in the DDR start to suffer from 200 MHz onwards. The frequency will be brought down to 100 MHz while maintaining the performance. This means that the frequency of the I/O buffer is doubled at the same time. The DDR2 rate of 100 MHz for the cells lets these deliver 4 bits per work cycle to the I/O buffer which works at a rate of 200 MHz. The bandwidth of a DDR2 at the rate of 100 MHz is therefore the double of a DDR at a rate of 200 MHz. The output based on a rate of 100 MHz is therefore ((100*4)*64)/8 = 3,200 Mo/s*

A RAM produced in 2005 consists of a chip of 4 Gbits (512 Mo) engraved on 0.10 micron. The new generation of circuits allows for the reduction of the price per unit (i.e. price per bit) by more than 60% in comparison to the chips of 128 Mbits. They allow an output of more than 10 gigabytes/s. The generation of DDR3 should bring about a bandwidth of more than 16 gigabytes/s.

In order to "write" a single bit of data into a new memory cell of the RAM, a transistor transfers around half a million electrons to a neighboring capacity. This minimal quantity of charge carriers is necessary to ensure the reliability of the saved information, i.e. to preserve the data and make it detectable in the field of ambient electric noise.

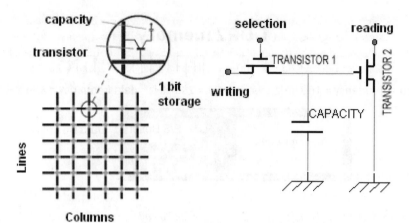

Figure 5.4a. *A DRAM stores a bit of information in the form of an electrical charge. It uses the capacity of certain semi-conductors (capacity of a wired grid of the MOS transistor)*

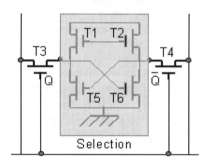

Figure 5.4b. *SRAM (static RAM) works differently from DRAM. Its content does not need to be refreshed periodically. One advantage of this kind of memory is its speed. On the other hand, it is much more expensive than DRAM and takes up four times more space than the latter given its limited density of integrating data: SRAM consists of a flip-flop. Every flip-flop consists of six transistors. The information remains saved as long as there is a different potential between the different terminals of the cell. This form of RAM also needs a permanent form of power supply. However, it uses an extremely low current of electricity which could be supplied by a small battery. DRAM, on the other hand, stores an electric charge in the capacitor which is affected by leaks and needs a refresh period, consuming higher amounts of energy*

The highest density in information storage is subject to the size of its capacities or very small capacitors whose dimensions are proportional to the quantity of charge carriers needed to save the data. A device which works by using less electrons to save a piece of information would therefore be even smaller and would consume even less energy.

With current technology, electrons are used each time information is "written" and therefore saved somewhere. The faster a computer is in these writing processes, the more energy it consumes. These are therefore not very compatible with today's equipment which is becoming more and more portable and therefore restrains the amount of energy that can be used. Furthermore, traditional DRAMs do not limit energy consumption to the process of "writing" a piece of information. In order to save a piece of information, an electric current needs to be injected into the circuit on a permanent basis. If the power supply fails, the information is lost.

The concept of a non-volatile RAM is the main objective for researchers in this domain. A single-electron memory is the ultimate aim of this technology. It would allow for an extremely high density of integration, a much faster access (less charge carriers lead to less inertia) and a very low energy consumption. In 1999 Hitachi announced the creation of a technology which was able to create memory chips of 8 Go with an access time of 10 nanoseconds. An example of this would be all of the sound effects of a film which could, in this case, be stored one a single chip. Since this technology is able to save the data when the computer is switched off, it could be a replacement for the hard disk.

This technology is called PLEDM (*Phase-state Low Electron-number Drive Memory*). It only uses 1,000 electrons to store one elementary bit, which is 500 times less than traditional DRAM as the information is not stored in a capacitor but in the gate of another transistor, in other words, an amplification cell which introduces a transistor at a low electric noise so the signal of 1,000 electrons is reinforced.

The amplification transistor is installed underneath the storage transistor. Both are installed in the same cell in order to integrate this circuit on an even higher level. The storage transistor uses an electronic shutter which closes the transistor in the case of a power supply failure so the electrons remain inside the cell. In fact, this new generation of RAM could combine the advantages of the currently used DRAM (high density of integration) and SRAM (non-volatile) in order to improve the performance of both technologies. Functioning in the same time as a RAM and as a miniaturized mass storage, this would have a major impact on the most significant domains in the digital world such as portable devices, telephones, audio/video players and recorders as well as PDAs, etc.

144 Nanocomputers and Swarm Intelligence

This technology is the starting point of the transition process towards a single-electron memory in which the presence or absence of a single electron in an atom would allow a single binary information to be saved. This miniaturization of memory cells will certainly take a different route from traditional transistors. In 2006, researchers working for IBM in Zurich managed to trigger a reaction of an organic molecule which was smaller than 1.5 nm (100 times smaller than current transistors) with an electric stimulation and therefore store and retrieve binary information from it. This synthetic molecule called BPDN-DT preserved its capacity of functioning as a capacitor over several hours and after more than 500 tests. This is currently a remarkable result for a monomolecular system.

The exact process of communication with this molecule still needs to be analyzed in great depth. The extremely high speed of communication, however, is already certain. Current tests revealed that the device reacts in less than 640 ms. The outer limits of this speed have not yet been established, which could theoretically allow for the production of memories that work at more than 1,000 GHz!

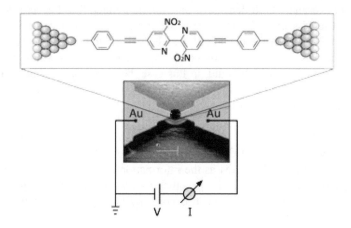

Figure 5.5. *The experimental circuit switching of an BPDN-DT molecule. The synthetic molecule is placed between two electrodes and under a Scanning Electron Microscope (SEM)[1]. These electrodes function as an extension which will eventually break down the reversible "nano-bridge" which forms the molecule that communicated with the circuit. This method is called mechanically controllable break junction (MCBJ). The electrodes are adjusted to a tolerance in the spectrum of the pictometer (10^{-12} m). In these dimensions, the Joule effect is a major obstacle in the experiment (source IBM)*

---

1 Scanning Electron Microscopy (SEM) is based on the principle of interactions between electrons and matter. A beam of electrons is shot towards the surface of the sample which in response sets certain particles free.

Elements of Technology for Information Systems   145

Another approach to future memory exists using semi-conductors. This approach does not use an electric charge to store information, but a magnetic field, or more precisely the phenomenon of spintronics observed in the individual components. It is called the tunnel magnetoresistance effect. This device is known as MRAM (*Magneto-resistive Random Access Memory*). This type of memory consists of a magnetic tunnel junction with a giant tunnel magnetoresistance effect (see section 5.2.2). It is associated with the majority of the advantages of memory based on semi-conductors currently available on the market, such as high speed when storing and reading data, a low energy consumption and being non-volatile. Because of its increased degree of integration (small sized cells and the possible reduction of the homothetic size) and its infinite cyclability[2], it has been the successor of flash memory which is rather slow in the writing process and has a high energy consumption. Its imperviousness to radiation, to name just one of the DRAM's qualities, will, in the medium term, lead to its application in the military domain as well as space travel.

IBM and Infineon were the forerunners of this new technology. Sony, Philips, Motorola, Toshiba and NEC have joined them as well. The first applications on a wide scale concern the generation of cell phones from 2006 onwards. This market could represent $2 billion up to 2008. By 2012, it is estimated to reach $12 billion.

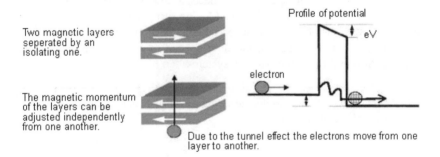

Figure 5.6. *The tunnel effect is designed to fit the quantum object's properties and may break the barrier of the potential. According to traditional mechanics, such a breakthrough is impossible: a magnetic tunnel junction (MTJ) consists of two ironmagnetic layers with different magnetic polarization which are separated by a fine layer of isolation. Electrons which are injected perpendicularly into the MTJ are polarized on their way by magnetic electrodes and permeate the isolating barrier with the tunnel effect. This tunnel transmission depends on the relative magnetization of the two electrodes to the interfaces with the tunnel barrier and the type of barrier. The resistance of such a junction can therefore take on two extreme states; one for a parallel configuration and the other for an orthogonal configuration: this phenomenon is known as tunnel magnetoresistance (TMR)*

---

2 The cyclability gives the possible number of writing and reading cycles, which characterizes the lifespan of a storage device.

In a "standard" layout of MRAM, the activation of the parts responsible for the writing of information is carried out with the help of minuscule magnetic fields induced by a lattice of semi-conductors which are subject to electric impulses.

Only the part in the memory which is situated at an intersection of two lines of selected current (i.e. lines/columns) turns around (the spin is inversed).

During the reading process, a low current is sent by a CMOS transistor to read the resistance of the junction. This layout brings certain limits with it. These are, for example, errors concerning the activation and the distribution of fields destined to inverse the spin in the memory's lattice, the limitation of the ultimate density due to three different levels of lines which move the transistor to the desired point within the memory and, last but not least, the high energy consumption.

5.2.2. Limits of magnetic data storage

More than five million terabytes[3] are stored each year on mass storage devices. This includes hard disks, optical devices such as CDs and DVDs and other magneto-optical devices. Magnetic hard disks make up two thirds of the entire market of storage devices. These storage devices need a few milliseconds to access the information and are much slower than RAM. The surface which is occupied by a bit of information on a disk has been reduced by a factor of 100,000 in the last 30 years. The capacities of data storage and the time needed to access them have improved and led to a drop in the prices of storage systems.

> "In the future we will ask ourselves about the era when man had libraries in which the books did not talk to each other."
>
> (Marvin Minsky)

---

3 The USA alone produces 40% of all new information that is stored. This is one-third of the worldwide amount of information produced by the press and the film industry. The USA creates 40% of all information stored on optical devices and 50% of all information stored on magnetic devices.

Elements of Technology for Information Systems   147

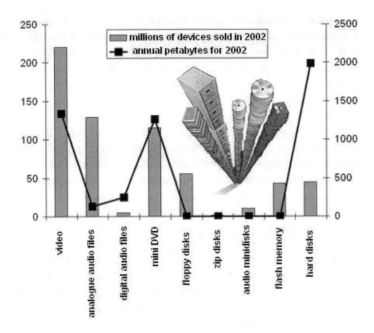

Figure 5.7. *In 2002 the press, film and music industries and all kinds of information systems created about 5 exabytes of data (5×10^{18}). In order to understand this figure it is important to know that very large libraries consisting of several million books such as the National Library of France or the Library of Congress would contain more than 100 terabytes if all their content was digitalized. 5 exabytes is therefore more than 40,000 of those libraries. 92% of all information produced in 2002 was stored on magnetic devices. Most of it is stored on magnetic hard disks, by far the most common device*

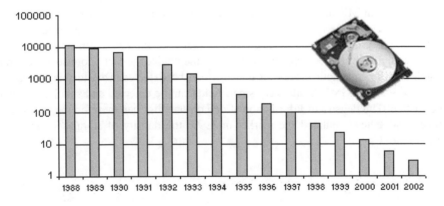

Figure 5.8. *Cost of a gigabyte on a hard disk (in $/Go), logarithmic level*

The race towards a higher memory capacity and speed of accessing the data is taking place in an incredible manner. The density of storage is increased by reducing the size of the magnetic mark left on the disk. This means that in order to increase the density, the size of the disk's read/write head needs to be reduced but its sensitivity needs to be increased. These improvements depend very much on the materials used.

A density of storage higher than 80 Gbits/in^2 can today be applied with the exploitation of GMR, which was mentioned in the previous chapter. GMR is a phenomenon that very rapidly decreases the resistance of certain materials when a magnetic field is applied to them. This phenomenon is referred to as "giant" since never before has such an effect been observed in the field of metals. The GMR phenomenon became very interesting for industrial and commercial ventures as it detects a very small magnetic field due to an altering electric current.

GMR disk read/write heads appeared on the market in 1997. Most modern computers are currently equipped with them. A GMR disk read/write head can in its structure be compared to magnetoresistive heads, which were used in previous generations of disks. The principle is based on detecting induced electric current through a coil which is subject to an electric field. MR heads consist of a receiver made up of NiFe in which the resistance changes according to the magnetic field which can be found on the surface of the hard disk. In the case of GMR heads, the writing process is always carried out by an inductive head. The reading process, however, relies on the quantum properties of an electron with two possible spins. When this spin runs parallel to the orientation of the magnetic field of the hard disk a weak resistance can be observed. If the spin, however, takes place in the opposite direction a strong resistance can be observed.

Changes in MR heads affect the electric resistance of an element placed near a magnetic field. However, GMR heads use extremely fine layers of material, so the reading and writing process is much more exact than the one given by MR heads. This is why they are described as "giant". As the device is much more sensitive and much smaller, magnetic traces on the disk can be detected. Due to this increased sensitivity of the GMR heads, very small fields can be detected and smaller bits can be written. The density of information stored on one disk can therefore be increased. Since GMR heads entered the market in 1997, the density of storage increased a hundredfold.

Elements of Technology for Information Systems   149

> The concept of a magnetic hard disk originated in the 1950s. IBM's engineers discovered that the read/write heads, if they were constructed appropriately, could work from underneath the surface of the rotating disk and read the data without establishing direct contact with the disk's surface.
>
> The first hard disk produced by IBM was called RAMAC *(Random Access Method of Accounting and Control)* and introduced in 1956. The device weighed nearly one ton and had a capacity of about 5 megabytes (2 Kbits/in^2) as well as a transfer rate of 8.8 Ko/s.
>
> In 1973, IBM presented the 3340, the predecessor of the modern hard disk, which for the first time was based on a sealed environment. Each unit consisted of separate modules, one stable and one movable, and each of them had a capacity of 30 Mo. This is why the disk was often called "30-30". Referring to the Winchester 30-30 gun, this hard disk is nicknamed the Winchester hard disk.

At present, we are reaching the limits of "traditional" GMR heads, i.e. GMR heads with the electric current running parallel to the layers. The next generation of disks will use other effects of spintronics such as TMR (see section 5.2.1) or GMR with a current that runs perpendicular to the layers.

Another approach to increase the density of storage uses a vertical orientation of the disk's magnetic dots so that they are perpendicular to the surface of the disk rather than "lying flat" towards the read/write heads. If the size of a magnetic dot decreases, the magnetic field also decreases more quickly. This phenomenon is known as auto-depolarization. A periodic refreshment of the content is required more frequently.

However, all these incremental improvements in storage devices share the same fate as expressed in Moore's Law. They will "hit a wall" which will manifest itself as an asymptote of their performances. If the density of storage on disks reaches a three-digit growth rate, this could not be more than 100 Gbits/in^2 (the very next generation). A development which exceeds these dimensions will require new technology.

Saving data with an ultra high density could be an alternative to the present system which is based on a different approach. Information would no longer be written on a medium such as the currently used hard disks, but on a magnetic contact of a length of several dozen nanometers. In this case, the density of storage could reach 200 Gbits/in^2 instead of today's 80 Gbits/in^2. Optimists imagine that this approach would allow the limits of magnetic storage to be pushed even further towards dimensions of TeraBits/in^2.

> In portable equipment, miniaturization is the essence for future generations of digital ubiquity. An increasing number of experts consider the relatively fragile hard disk, in terms of miniaturization and the fact that it is a rotating mechanical system, which is far from being ideal, even in the short term. Samsung, amongst others, drew its conclusions from this fact: solid state disks (SSD) replaced the traditional hard disk drive (HDD) with a flash memory. The Korean producer wanted to take over 20% of the storage market by 2008 with this technology, producing hundreds of millions of them. According to Samsung, the market for flash memory chips and the NAND technology of 8 and 16 Gbit was booming by over 100% in 2006. The company will therefore alter its production by 70 nm, changing from the SLC (*single level cell*) technology to MLC (*multi-level cell*[4]). The volume of SLC will remain stable but its size will decrease to 55 nm.

5.2.3. *Holographic memory*

The principle of holograms has been around since 1947 and the idea of storing data in them developed in the years following their discovery. The first devices to store data for information systems which experimented with this idea are, however, very recent. Since the mid-1990s, big research laboratories such as IBM, Lucent Technologies and more recently Imation have intensified their research in this field. They started mainly because of the incentives offered by DARPA (Defense Advanced Research Project Agency).

Traditional technologies reached a standstill and the developments in the density of storage that concerns them are already a thing of the past. Around one gigabyte of information can be stored on a CD, while a DVD holds around 17 gigabytes. Tests carried out on the magnetic disks mentioned above show that they can store around several hundred gigabytes and are approaching the terabyte. However, the storage requirements of our future computers in the next five years greatly exceed what technology can provide. Amongst all completely new approaches that could be exploited on an industrial level, holographic memory could be a solution.

Holographic memory is an optical storage device that allows a density of saving data as well as a speed of accessing it which is much higher than in traditional devices (i.e. magnetic hard disks). A holographic image is encoded in the form of a very large block of data and saved in one single operation. Reading the data is carried out as a parallel operation, i.e. a block of data can be read in one single

---

4 MLC memories consist of two bits of stored data per cell. Therefore, four possible states exit: erased (11), two-thirds (10), one-third (01) or programmed (00). The more complex layout of MLCs is a disadvantage when comparing it to SLCs.

operation. These blocks of data are holographically stored on photosensitive material, a small crystal of lithium niobate (LiNbO$_3$), the researchers' favorite material for experimental holographic storage devices. To read and write these blocks, analog laser devices, which are similar to those of molecular organic memories, are used.

In theory, the density could approach terabyte/cm^3. Practically speaking, these new devices allow a density close to 10 gigabytes/cm^3, which makes it impossible to compare them to even the most advanced magnetic devices. This system is marked by a high increase in storage. It is capable of storing and extracting entire pages of data (each made up of several dozen megabytes) in one single operation without using a single mechanically movable piece.

How does holographic memory work? The hologram is created by the interference of a beam of light called an object beam and a second beam of light called a reference beam on a photosensitive material. The object beam first passes through a Spatial Light Modulator (SLM) that consists of a mask or a screen of liquid crystals. This mask can cover certain points and lets the light of the object beam affect only certain areas. This type of grid leaves black and white squares which symbolize the digits 0 and 1 which encode the information to be saved. By using mirrors, the reference beam is directed into the inside of the photosensitive crystal where it will then be met by the object beam. The "meeting" of the two beams leads to the creation of an "image of interferences" which modifies physio-chemical properties of the crystal's molecules, in particular, those of the refractive index. Every page of information is then written into the inside of the crystal under the form of a hologram. In order to read the information, it is sufficient to light the crystal with a reference beam. As the "image of interference" was previously stored, when exposed to the original beam it leads to a diffraction of light which represents the other beam and therefore creates the page of data. Now the image only needs to be read by a charge-coupled device (CCD) similar to those found in camcorders.

One crystal can contain several "images of interference" which all relate to a page of data, for example, an entire book. To select pages, the wavelength of the laser of the angle between the object beam and the reference beam is altered.

152   Nanocomputers and Swarm Intelligence

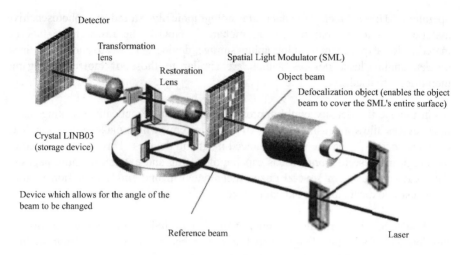

Figure 5.9. *The source, a laser, is divided into two components, the reference beam and the object beam. The object beam carries the necessary information. First of all, the object beam is devocalized with the help of an optic device, a baffle, in order to illuminate the SLM's entire surface. The SLM consists of a lattice of liquid crystals (LCD: liquid crystal display) which functions as a mask for the object beam and allows certain points to be covered, and lets the beam pass at other points creating a cross-ruling of black and lit up pixels. The two light beams penetrate the photosensitive crystal and interact with each other in the inside which leads to the creation of the "image of interference", which is stored as a group of variations of the refractive index in the inside of this crystal. In order to read the data, this crystal is lit up by the reference beam which interacts with the interference factor that is written into the image of the crystal and reproduces the page as a cross-ruling grid where black and white squares (pixels) respectively represent the 0 and 1 of the stored information. This information is transferred to a detecting device, the CCD. During the reading process, the angle of the reference beam needs to be the same as it was during the writing process, the tolerance is less than one degree. This weakness in the process is at the same time one of its strengths as it allows for an extremely high density of pages of data that can be saved inside a crystal. The smallest variation of the angle corresponds to a new page of data*

However, the process of reading and writing and therefore saving data is imperfect due to the electric field that stems from the reference beam. The data therefore needs to be refreshed on a regular basis in order to be maintained.

The holographic memory is therefore unable to replace read only memories (ROM), i.e. DVD players and other devices used as archives for data.

Polymers seem to have some advantage due to the form of crystals used in current experimental devices. Bayer in Leverkusen, Germany is working on photo-addressable polymers (PAP) which could be used more easily than the crystals,

Elements of Technology for Information Systems   153

especially when it comes to industrial mass production. Bayer predicts commercial usage in the next two years.

Traditional mass storage devices have to extract the information from the medium before being able to consult it and compare it to the requested operations (this is the IT toolbox needed for production). In the case of holographic memory, the data does not need to be extracted to proceed to the operation that compares it to other sources.

The selection of data according to given criteria in data warehouses or datamarts consists of very long procedures that "kill" content management systems. The time needed to carry out a search is now much shorter.

This property is the biggest commercial advantage of this device for the next five years as it can be applied to systems which lie at the heart of large companies such as ERP, CRM or Business Intelligence. The first systems that will enter the market will be destined for the storage of data in companies.

A market study stated that the first applications will offer a capacity of 150 gigabytes on a removable disk with a format of 5 1/4". The speed of transfer is estimated to be at 20 megabytes/s for the writing and 40 megabytes/s for the reading process. However, this technology is not necessarily limited to disk type formats, but could be reduced to every 3D immobile device.

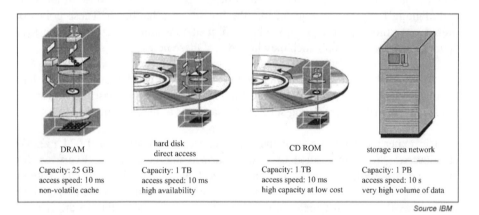

Source IBM

Figure 5.10. *Four possible scenarios for the development of holographic storage devices*

5.2.4. *The technology of AFM memories*

Atomic force microscopes allow for the exploration of all possible types of surfaces by only using the molecular force that the surface of the sample applies to the tip of the microscope (see section 3.1). The tip is in contact with the atoms of the surface and the vertical movement of the probe when it scans the sample's surface horizontally. The atoms are detected optically. If the tip is covered with some form of metal, the variation of the magnetic field emitted by the probe, which swings from side to side at a high frequency, can be converted into images, i.e. meaningful data.

The technology of AFM (*Atomic Force Microscopy*) memory exploits this principle with the aim of increasing the performance of two-dimensional storage.

The concept stems from IBM's laboratories in Zurich which already possess the STM and the AFM. This form of memory could be described as a nanoscopic version of the mechanical data storage used in the first computers, i.e. punched cards. However, this time the holes are "nanoholes". Furthermore, the cards of the nanotechnological generation can hold more than one billion holes per mm^2. Theoretically, a density of storage of about 300 gigabytes/in^2 is possible. Tests have so far been limited to a density of 64 gigabytes/in^2.

The Millipede project developed by IBM in Zurich consists of developing a 3 mm squared device made up of billions of points derived from those of an AFM and aligning them on this square, like the bristles of a hairbrush. These points or probes only move on a square of plastic film of few mm^2 in size in which they "punch" nanoholes, or on which they later detect these holes.

"Millipede" indeed refers to the creature and its ability to move all of its legs simultaneously. The system allows for reading and writing processes to be carried out simultaneously by activating several thousand tips at the same time.

The experimental setup contains a 3 mm by 3 mm array of 1,024 (32x32) cantilevers, which are created by silicon surface micromachining. Every tip (or probe, i.e. the millipede "legs") has its own working zone, and all tips can write or read at the same time. In order for every tip to be able to read and write all positions in its zone, the grid on which the tips are affixed moves according to the X or Y axis, which are determined by the storage support device. At a particular point in time every tip will address the same position within its respective working zone. These movements of X and Y on the matrix allow the encoding device to read and write pages of data structures in tables (rows and columns) of binary data. Every pair in the position of (X,Y) corresponds to one page of data. An activation mechanism allows the selection of a pages (position (X,Y)) on the grid and ensures the contact between the two points and support of polymers (position Z). A page of 1,024 bits is therefore read or written in one single operation, which is a step ahead of the relatively slow process of reading and writing of the best performing magnetic systems which are limited due to thermo inertia. The devices authorize very high transfer rates for two-dimensional memory approaching the 10 Mbps mark.

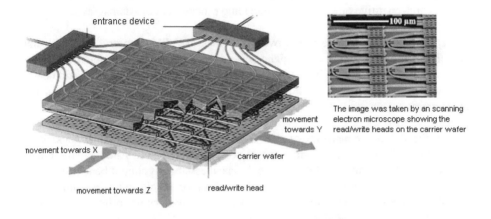

Figure 5.11. *The technology of AFM memory, IBM's Millipede project*

The storage support device is made up of three layers: the first layer consists of PMMA (*PolyMethylMethAcrylate*) and corresponds to the support into which the hole is engraved. The second layer consists of Su8, a photosensitive resin which is there to stop the engraving process and directs heat from the tip to the third layer. The third layer is made up of silicon in order to absorb the residual heat. The second layer acts as a softer penetration stop that avoids tip wear but remains thermally stable (e.g. better quality of the holes and higher resistance towards heat, which is essential for the reading process).

The legs consist of tiny built-in girders whose ends remain free and are the mono-atomic tips of an AFM. Every leg is about 50 microns long and consists of two parts which allow for the circulation of two currents of different temperatures. The first current allows for a temperature of 350°C which is used for reading the data. When the second current is added the temperature rises to about 400°C, the temperature needed for the writing process. This tip penetrates the support and is stopped by the layer which is made up of Su8 and disperses the heat. As a convention, a hole corresponds to binary 1 and the absence of a hole to 0.

For the reading process, the tips are heated to 350°C and the grid moves. Unless the tip encounters a hole, heat is dispersed and the strong decrease in the tension of the leg is interpreted as 1. Holes are detected when there is no strong decrease in the tension. These intervals are interpreted as 0.

One of the main challenges lies in the material used to create the support in order to allow for repetitive operations of the reading process without damaging it.

IBM's researchers have tested two prototypes of ATM memory. The first one comprises five rows of five legs. The second one was previously mentioned and consists of 32 x 32 legs. These tests were mainly focused on different parameters such as the space between the holes, their diameter or the appropriate material for the support.

On the devices consisting of 32^2 disposing of 80% of functional legs, it is possible to write information with a density of 30 gigabytes/in². Another experiment showed that the density of 200 gigabytes/in² is accessible. It should be kept in mind that currently used hard disks do not exceed a limit of 80 gigabytes/in². During the test, the device was limited to one leg, which of course does not reflect reality. The density reached was several terabytes/in².

IBM believes that this type of device could be produced on a large scale and be commercialized in the short term, i.e. in the next few years. These devices could find their first applications in digital cameras and camcorders. They would be a strong competitor for flash memory because of their high capacity. However, some obstacles remain in the endeavor for mass production. These problems lie in the area of the stability of bits. The process of erasing and rewriting data also requires improvement. Furthermore, the density of storage and the number of legs need to be heavily increased in order to equip the first commercial applications and create a commercial advantage. Millipede 64^2 memories, consisting of 64 rows and 64 legs, are currently being tested. They are equipped with smaller legs and use a new schema of multiplexing in the activation device. Their energy consumption has been reduced, and in comparison to previous models like the Millipede 32^2 they offer a higher speed of data transfer.

5.2.5. *Molecular memory*

The idea of using organic and biological molecules taken from the world of living organisms, such as proteins, to encode, manipulate and restore information, has been explained in section 3.6.3.

> A protein is a nanomachine whose molecular structure has been selected during evolution in order to fulfill a specific biological task.

We have seen how molecular memory use the photo cycle, which means that the structure of a molecule changes when it reacts to light. Here, photosensitive material is used in the cycle to change the protein from a stable state to another state, such as a messenger which encodes and then restores information. This can be compared to traditional RAM. The major property of this technology lies in the protein's capacity to take on different three-dimensional forms and organize itself in the structure of a cubic matrix within a gel of polymers. Contrary to traditional memories that only use two dimensions to store information, molecular memories use three dimensions. In other words, the density of storage and the time required to access information (this device is based on parallel operations in terms of accessing data) have greatly improved. Their capacity lies within about 1 terabyte/cm^3 and a speed of access of 10 Mbits per second.

On the basis of this technology, Robert Birge produced the prototype of a hybrid computer which consists of four main boards on which storage and calculating units are divided and specialized. The board within the main unit only contains circuits made up of semi-conductors. Two boards are equipped with memory based on proteins (digicubes based on bacteriorhodopsin), which are capable of storing a maximum of 40 giga words. These boards allow random access without any mechanical part being used in the device as is the case of RAM within currently used computers. The other board allows for a more permanent and more economic storage. The fourth board contains an associative memory made up of fine layers of bacteriorhodopsin. The main advantage of such a computer lies in its exceptionally high storage of around one tera-word (10^{12} words). This capacity is particularly important for complex scientific simulations or cognitive systems based on programming techniques that lead to artificial intelligence.

5.3. Batteries and other forms of power supply

The on-going progress in processing speed, storage capacities, communication between different machines and bandwidth has reduced the size of computer systems. The era of mobility has begun. Whether it is cell phones, PDAs, laptops, handheld game consoles, or even more innovative equipment such as *wearable computing* (i.e. computerized clothing), all of these rely heavily on mobile energy sources. Who is not aware of the weaknesses in current power supplies after experiencing a flat battery in their cell phone or laptop?

The technology of batteries has evolved rather slowly compared to that of semiconductors and mass storage devices. The rate of innovation only lies at 5 to 7% per year since the introduction of lithium technologies. Power supply, which has so far remained of rather little interest, has become a major challenge in the fields of portability and nomadic systems.

Current innovations of battery components are mainly being carried out in the domain of chemistry rather than in the domain of electronics. Today's batteries work in the following way: they create a link between two chemical reactions, an oxidation process which generates negatively charged ions or electrons and a reduction process that consumes those electrons. Furthermore, a "path" is created which the electrons use to circulate from the negative terminal to the positive one. These "terminals" are a positive and a negative electrode put into an electrolyte solution which allows the current to circulate in the inside of the battery. The amount of energy in the battery is limited to the quantity of energy stored in it under the form of the materials that participate in the reaction. When the electrons circulate between the negative electrode (anode) and the positive electrode (cathode), the battery generates energy while simultaneously discharging itself.

Improvement in this process currently relies heavily on the widely known technology of nickel-cadmium batteries (NiCd) and nickel metal hydride batteries (NiMH). New emerging technologies such as lithium-ion, lithium-polymer and zinc-air batteries are the focus of research. Progress in the field of nomadic systems and the increasing commercial demand for them is stimulating the development of conventional batteries and also making way for alternative solutions that do not have to rely on the progress made by the chemical industry. These are systems using kinetic energy to recharge batteries.

For ubiquitous computing and the generation of objects that communicate with each other, new approaches are currently being tested. These are power supplies generated from the world of living organisms, microcombustion engines, nuclear batteries and other devices that are very different from traditional batteries. Scientists are currently carrying out experiments on solutions capable of supplying micro and in the future nano-objects with energy that stems from living organisms, such as energy taken from spinach, sugar, slugs or flies. These new sources of energy use ATP (Adenosine TriPhosphate) molecules which store phosphates of high energy in order to store energy just like in the muscles of living organisms.

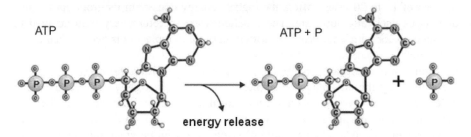

Figure 5.12. *Because of the presence of bonds which are rich in energy (bonds that link groups of phosphates are anhydride acid bonds), living beings use ATP to provide energy for chemical reactions that require them. The ATP molecule consists of three different types of units: a nitrogenous base called adenin, a 5 carbon atom of the pentose sugar called ribose and three groups of phosphates (H3PO4). The chemical bond between the second and the third phosphate can easily be created and also be easily detached. When the third phosphate detaches itself from the rest of the molecule, energy is released.*

The miniaturization of ambient molecular objects is not only limited to electronic circuits. To miniaturize any kind of technical object down to the micro or nanoscopic level, all technical functions need to be taken into account. This is more than just the integration of proteins and integrated circuits. Mechanical devices and the power supply that is needed for objects that are too small to be recharged on a regular basis need to be considered. Two approaches to these power supplies are currently being researched. One is embedded micro-batteries which rely on the conversion of chemical energy by a mechanical force. This process relies particularly on one protein, kinesin, which is present in all cells that possess a nucleus (eukaryote cells). This protein transforms chemical energy of the ATP molecule, which can for example be found in muscles, into mechanical energy

5.3.1. *Lithium-ion and lithium-polymer batteries*

The lithium-ion battery is currently used in the majority of portable equipment from laptops to cell phones as it is relatively small and light. Currently, this battery supplies the strongest specific energy (energy/mass) and is of the highest density of energy (energy/volume). It consists of a carbon anode and a lithium cathode, i.e. oxide transition metal whose valence electrons can be found on two layers instead of just one. An organic solution is used as the electrolyte. The disadvantage of this kind of electrolyte is that it is very flammable. In "traditional" batteries, the tension between the electrolytic poles lies at about one to two volts per cell. The lithium-ion battery generates up to four volts per cell which explains its excellent performance in terms of specific energy. For a large number of applications, the lifespan of this battery may reach seven to 10 years. The lithium battery using thionylchlorid has a lifespan of 15 to 20 years, offers the highest energy density in its group and it also discharges itself at a low rate. These batteries are therefore very well adapted to applications for which access to the battery in order to replace it is very difficult.

Lithium-polymer batteries are the next generation of lithium-ion batteries. They are marked by the advantages of lithium, such as being lightweight. At the same time they do not have the inconveniences of the lithium-ion batteries (e.g. they are very flammable, discharge themselves more quickly, etc.). In the case of lithium-polymer batteries, instead of a liquid electrolyte, a solid polymer substrate is used. The lithium ions are incorporated into this substrate so it can be easily worked on. This type of battery is environmentally friendly and is lighter and more reliable; it does not require a metal case which usually harbors the liquid components of batteries of the lithium-ion group. As they do not need the metal case, lithium-polymer batteries can take the shape of very thin pieces of paper, making way for a much wider field of applications.

5.3.2. *Zinc-air batteries*

The zinc-air battery has one very original property: it uses the oxygen of the air surrounding it. The oxygen therefore does not need to be stored in the electrolyte. This group of batteries consists of a zinc cathode and a graphite anode. It uses the atmospheric oxygen as an electrolyte. These batteries offer a much higher density of energy than the majority of alkaline batteries. They are equipped with functions that block the circulation of air during the charging process and when in a passive state. These functions also ensure ventilation of the cells during the discharging process which allows for an increase in energy and limits auto-discharging in the passive state. The lifespan of this type of battery is generally five times longer than that of other products on the market.

By using oxygen as the electrolyte, this technology allows for a smaller battery. On the other hand, air also contains elements that may damage the battery: steam from water and carbon dioxide are the two main ones. Variations in the air's humidity can significantly affect performance. Furthermore, carbon dioxide forms carbon crystals in the cathode that are supposed to block the circulation of air which leads to less energy being delivered and also limits the cell's lifespan. The layout of the tube through which the air circulates is essential, it may be long tubes with a rolled-steel equipped with sections of hydrophobic membranes. This system keeps most of the steam from the water in the air outside and therefore allows for an efficiently working battery even under the worst conditions.

Elements of Technology for Information Systems   161

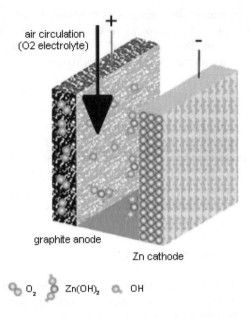

Figure 5.13. *The zinc-air battery uses the oxygen in the air surrounding it. The oxygen therefore does not need to be stored in the electrolyte*

5.3.3. *Micro-batteries*

The implementation of embedded microsystems whose capacitors, processing units, activators or transmission systems are integrated into areas which are difficult to access (such as the engine of a vehicle, on a moving part or inside the human body) is indeed very difficult. This is why embedded microsystems need their own source of energy. Furthermore, the microsystem and its power supply cannot exceed the sub-millimetric dimension.

Developing a "traditional" battery on this level is, of course, impossible.

A new generation of batteries, the size of the width of a hair, with a density of energy which is 100 times higher than that of "traditional" batteries, is currently being developed. These batteries are constantly recharged by devices that recover energy through magnetic vibration (see section 5.3.5), offering long term independence from a stable power supply to embedded microsystems.

The concept's originality lies in its production and the research of new materials, which were used as electrodes in the form of very fine layers. The components V_2O_5

and $Fe_xV_2O_5$ were used as positive electrodes in lithium batteries. Instead of putting the battery together piece by piece, as is the case for traditional batteries, the technology derived from the production of semi-conductors is applied. This technology relies heavily on the process of photolithography.

Such a process is much simpler than the traditional production of batteries and can be easily used on the micron level. Furthermore, it also allows for the usage of materials of a remarkably high level of purity.

This kind of power supply will be used in the first applications of embedded microsystems that are implanted into the human body. These systems will, for example, be able to regulate the level of insulin in a diabetic patient. These systems need to be perfectly autonomous and their lifespan needs to be several years.

5.3.4. *Micro-batteries using nuclear energy*

Atomic engineers at the University of Wisconsin, Madison have explored the possibilities of producing minuscule batteries which exploit the decrease in the number of radioactive nuclei of substances such as polonium or curium to produce electric energy. They can therefore be used as a weak power supply. These devices, which are similar to MEMS (*Micro Electro-Mechanical Systems*; see section 5.6), measure between 60 and 70 microns.

These micro-batteries are very light and deliver enough energy to function as a power supply for nanodevices.

The energy released through the decrease in the number of radioactive nuclei is transformed into electric energy, which is restricted to the area of the device that contains the processor and the micro-battery that supplies it with energy.

The decrease in the number of radioactive nuclei releases electrons that shoot the negative pole of the battery and create a difference in the potential which then leads to an electric current that supplies the micro-equipment with electricity. To avoid the electrons reaching the negative/positive pole in a neutral state, a "void" is created between the two poles and the electrons are forced to pass the radioactive material at the negative pole through a minuscule conductor line. Ambient energy that is released by the radioactive substance does not entirely consist of charged particles (electrons that will create electric energy) but most of it is released under the form of heat which could be used as a form of power supply by the micro-equipment.

The energy supplied by this kind of device is measured in microwatts or in milliwatts. This type of source can supply energy for small portable devices, chemical receptors and micro-laboratories (lab-on-chip). Scientists are already planning the production of microsensors based on this type of power supply. These sensors will be added to the oil that ensures the smooth running of heavy machinery. In this way, they carry out a local micro-control for an improved maintenance of the machinery.

5.3.5. *Recharging batteries with the help of kinetic energy*

This principle is based on the conversion of kinetic energy into electrical energy in order to recharge a battery, for example, the alternator of a car through which connection with the combustion engine constantly charges the battery. A large number of research projects are being carried out and some of them are close to being introduced to the market. Compaq has put a patent on a device that uses the mechanical energy of typing on a keyboard to recharge the battery of the laptop in use. This principle is very simple and consists of connecting every key with a small magnet around which an inductor is wrapped. Each time one key is hit the magnet creates a magnetic field which is converted into electric current that then charges a capacitor. When the level of energy in the capacitor is high enough, the energy is stored and transferred to the battery.

The concept of large wireless microsystems that exchange information relies on minuscule systems which are autonomous when it comes to power supply. However, this autonomy is limited by the spread of a high number of batteries used in a certain network of objects (see section 6.4). They are a major source of pollution. The solution might lie in recovering energy (mechanical, chemical, thermal or radiant) in the surrounding area of the systems in order to make them autonomous during their lifespan. Recovering energy through mechanical vibration, used, for example, on moving vehicles or in industrial machinery, is currently one of the most researched solutions in this field. With the help of mechanical converters using silicon technology whose frequency lies around some 10 Hz and vibrates with the amplitude of some hundred microns, a device may have the power of around one microwatt per gram of mass. The power the device may use is around several miliwatts and therefore compatible with the energy the equipment requires to function.

5.4. New peripheral devices and interfaces between humans and machines

The concept of pervasive technology characterizes a new generation of information systems, relies on two principles and goes much further than miniaturization or increases processing and storage capacities. These are the "democratization" of devices in terms of the price and the development of a better interaction between humans and machines.

Human computer interaction (HCI) tends to erase the notion of an interface by merging reality with the world of machines. As technology is increasingly used in our daily lives, such as receptors or other devices that are placed underneath our skin, or the use of electricity produced by the human body which is used as a power supply for appliances that lie on our skin, progress in this field will allow for an osmosis of technology and the human body. The development of these interfaces often merges current scientific disciplines in research and development. The more our environment is interlinked, the more we need to adapt access to fairly complex information.

If traditional keyboards and screens of the new generations remain the principal means of information exchange between our world and that of information systems, new possibilities for the machines to obtain information should be created.

On the one hand, these new approaches allow computers to obtain information and orders from the user in a more natural and anthropomorphic way. Peripheral devices allowing for the entering of data will be an extension of the human body. Some of these systems which are already operational have even been commercialized, while others are still at the prototype stage. They mainly include functions such as eye tracking, recognition of gestures (handwriting recognition has been applied to the keyboards of certain devices for several years now) and voice recognition, which is well known and a rather basic device. Highly pervasive systems or "diluted" information systems can use other channels of information exchange. This exchange is more than the communication with the macroscopic world of humans (voice and gestures) and the world of machines (receptor of position, temperature, pressure, etc.). The desired information is of a microscopic nature from minuscule systems often integrated on just one chip which can hardly be distinguished from the surroundings it controls.

Another aspect of the interface between humans and computers of the new generation is the concept of recovering and visualization of data. "Electronic paper" is a device that possesses the characteristics of traditional paper; it is very fine, light and flexible with clear contrasts once something is written on it. It is also able to maintain the visualization of information when it is not connected to a power supply. Furthermore, it can be reused i.e. the content can be erased and new content can be uploaded. This is one of the new visualization devices which has recently entered the market. If an antenna is integrated into a sheet of paper it can receive data via radio waves from a newspaper for example, and visualize the content with help of the electronic ink that can be found in the paper.

Apart from this example, other visualization devices of multimedia content are being developed to fit all possible situations in daily life. In the short term, successors of LCD screens will be developed which do not need their own source of light to ensure the visualization process. This weakness is a disadvantage of the LCD screens that took over from the "antique" CRT (Cathod Ray Tube) screens. LCD screens still use too much energy, are too heavy, too thick and too fragile for large-scale distribution. Research is being carried out into devices which can generate their own source of light by recycling the light available in their surrounding area in order to ensure their own light source.

> In the case of laptops, their screens use most of the electrical energy. The goal of ubiquity and pervasive systems is based on a massive diffusion of mobile and portable devices. However, it is currently being hindered by the visualization multimedia content.

OLED screens, for example, emit visible light from any angle and allow for more realistic and clearer images. They also consume less energy than LCD screens and are much cheaper in production. As they are very slim, they are compatible with the idea of electronic paper and the mobile usage which characterizes the new generation of our information systems.

As the subject is very broad, there will be no detailed elaboration about new technologies in the fields of video and photography as it will take us too far away from the original topic. Some of the emerging fields will still be mentioned in order to give an overview of the new generation of computers including the visualization of information for the user.

5.4.1. *Automatic speech recognition*

The interface between humans and computers has evolved since the appearance of the mouse and more recently with touch screens. All interfaces use mechanical force that is applied by the users' hands and fingers. Since the mid-1990s, automatic speech recognition (ASR) has been developed as an extremely efficient alternative for communicating more naturally with information systems of embedded systems.

The first application took place in the Bell laboratories in 1952 when a team of researchers developed a wired logical system that allowed for a computer to identify 10 decimal numbers and the words *YES* and *NO*. The device imposed an extremely slow rhythm of speaking and did not accept the slightest variation in the articulation of words. Pauses of a tenth of a second were taken between each word and were used as a separator. The first systems could only be used by one speaker and often required a long learning process before the machine could process the data delivered by the user.

Even today, machines can process several hundred words but the learning process is still long as the machine needs to save a sample of all those words it is supposed to "understand". Systems that can process an even larger vocabulary, i.e. several thousand to several hundred thousand words, ask the speaker to recite a group of sentences which will be analyzed by the interface in order to establish the parameters of a model.

Today, systems which allow for several speakers to use them only exist with a very limited vocabulary (some words and 10 numbers). They are used for voice mail applications or for dialing numbers orally.

At present, not every field is experiencing progress due to this new generation of technology. It is used for voicemail in telephones, dialing numbers without using our hands, in more complex transactions in ERP applications and, last but not least, as a form of signature. Our voice is, of course, much easier to imitate than digital fingerprints or eye scanning.

These applications allow for the installation of a natural interface with the capacity to dictate directly to a computer that is able to understand up to several hundred thousand words and whose vocabulary can be extended by the user's own specialized vocabulary. Even when already using both hands, for example when driving, the user is still able to do business, the same way as with a real assistant: the user can dictate a text or give orders, and also ask for certain research to be carried out or the opening of a folder. This is the case for a highly elaborated interface such as structured query language (SQL) or any other fourth generation language. In warehouses and on logistical platforms, voice recognition is increasingly used to carry out transactions that are relative to supply chain management, i.e. the system that controls entire processes from the warehouse to the ERP. The forklift truck operators can carry out these operations while driving a forklift truck. Before the introduction of voice recognition they were forced to stop driving to carry out administrative tasks. Vocal recognition is an interface whose potential of productivity in the administrative sector is far from being used to its full potential.

The process of voice recognition consists of three main functions: analysis of the received signal (*front-end*), learning process (*training*) and recognition (*back-end*).

Our voice is a sound wave produced by the mechanical vibration which is carried out due to the flexibility of air in the form of analog. The vocal signal is caught by a microphone and digitalized with the help of an analog/digital converter. The human voice consists of a multitude of sounds, most of them repetitive. The signal is

Elements of Technology for Information Systems   167

generally compressed in order to reduce the time of processing and storing it in the memory. High performance applications on the market use highly reliable microphones and DSPs (see section 4.2.3) to ensure the high quality of the received signals, quickly produce samples and easily separate all other noise from the vocal signal. Standard applications use the PC's own microphone and sound card.

The first step consists of creating parameters for each signal, i.e. producing a characteristic sample of a sound which is then used to recognize it later on. There are several methods to do so, the most widely used ones are spectroscopic or identification based.

Spectral methods bring the sound down to its individual frequency without identifying its structure. In fact, the sound superimposes several sinusoid waves. By using an algorithm known as the Fast Fourier Transform (FFT[5]), different frequencies which make up a sound are isolated. By repeating this process several times a characteristic sample of the sound can be obtained.

Another approach called the identification method is based on the knowledge of how sounds are produced, for example, how the air passes the vocal cords. The buccal artery consists of a cylindrical tube of an altering diameter. Adjusting the parameters of this model allows for the determination of the transfer function at all times. The resonance frequency of the air passing the vocal cords can be captured. These resonance frequencies correspond to the maximum energy of the spectrum. By repeating this process a sample of the sound is created.

The signal is then sub-divided into workable units. The identification of long units such as syllables or words is complex but allows for easy recognition. Basic sounds would be much easier to identify but are not very reliable in the field of actual voice recognition. The compromise between these two is to take the smallest linguistic units which do not have meaning themselves but change meaning when used in different positions. In French, for example, there are 36 of these units, out of which 16 are vowels and 20 consonants. The French word "dans" (inside) consists of *[d]* and *[ã]*. The identification of segments is based on phonetic and linguistic limits of language (e.g. articulatory phonetics, the sounds of a particular language, phonetics, intonation, syntax, semantics, etc.).

The training function enables the program to create a link between elementary units and the act of speech with its lexical elements. This link uses statistic modelization such as the Hidden Markov Models (HMM) or an Artificial Neural

---

[5] The FFT is a mathematical tool used for processing signals. It associates a spectrum of a frequency with a non-periodic function.

Network (ANN). To create a dictionary that contains the acoustic references, the interlocutor pronounces the entire vocabulary used several times.

> An Artificial Neural Network (ANN) is a model of data processing that directly links to artificial intelligence as it is inspired by the functioning of neurons, i.e. the way our brain cells process, learn and store information. The main applications of neural networks are the optimization and the learning process.
>
> A neural network consists of a large number of basic cells that are interconnected, i.e. a large number of parallel calculating units. An elementary cell has a limited number of entrance and exit gates and can manipulate binary digits and numbers. Binary digits are represented by 0 and 1.
>
> This kind of network consists of simple processing/calculating cells (neurons) which are interconnected through bonds affected by weight or a weighting system. These bonds (or channels) enable each cell to send and receive signals to and from other cells in the network. Every one of these bonds is assigned to a weight which determines its impact on the cells it connects to. Every neuron has an entry gate that allows it to receive information from three different types of sources: other neurons, the activation function which introduces a non-linearity to the working of the neuron and the exit function which states that the result of the function carried out is applied to the neuron itself. In a neural network with N cells in the first layer $C(1)..., C(N)$ and a weight of N assigned to the bonds $w(1)...,w(N)$, the entrance gate of a cell in the second layer is generally the sum of the weight of bonds connected to the preceding neurons:
>
> $$X = w(1)*C(1) + w(2)*C(2) + ... + w(N)*C(N)$$
>
> Neurons or cells that process information are able to learn by adjusting the weight of their connections. The learning process consists of determining the optimal weight of different bonds by using a sample. The method which is most commonly used is known as retro-propagation: using the values applied to the cells on the entrance gate and with the errors obtained at the exit gate (delta), the weight is corrected according to the weighting system. This cycle is repeated until the rate of errors of the network stops increasing. If this process is carried out for longer than that, the network's performance decreases. This phenomenon is known as an "over trained" network.
>
> In the field of voice recognition, the neural networks are largely used in the isolated recognition of words, syllables and phonemes. The increasing demand in the field of continuous speech acts (i.e. a speech, sentences or words that are linked to each other and cannot be separated) progressively increases the usage of neural networks treating the continuous flow of the voice's signal in real time.

Voice recognition consists of recreating (back-end function) the most likely text. In order to do so, the machine ensures the concatenation of the elementary segments of the speech act in the form of a target text. Two approaches, one global the other analytical, allow for the recognition and comprehension of words.

In the global approach, the word is the basic unit and is considered as a whole entity which should not be sub-divided into smaller units. The system has an acoustic sample of every word and can therefore identify them. In order to do so, every word is pronounced at least once in the learning process and allows for the storage of this knowledge. The global approach avoids co-articulation, i.e. the assimilation of the place of articulation of one speech sound to that of an adjacent speech sound and leads to the sounds having an impact on each other within a word. However, this approach limits the vocabulary and the number of interlocutors. In fact, the recognition of isolated words requires a memory of considerable size to store all acoustic samples of all words in the vocabulary and an enormous processing capacity to compare an unknown word to all other words in the dictionary. Furthermore, it would be unrealistic to expect the user to dictate the entire vocabulary.

The analytical approach uses the linguistic structure of words. The basic units which are recognized are phonemes or elementary syllables. The analytical approach has a richer vocabulary as it is sufficient to record and save all principal characteristics of the basic units in the machine's memory.

Today, the demand is constantly increasing in the field of systems that allow the user to speak continually, i.e. naturally just as if the user was speaking to another person. Current voice recognition systems allow the user to dictate naturally and the words appear on screen with a delay of three to five seconds. Modern processors work at more than 1.5 GHz and have a memory of more than 1 megabyte, so this delay can be reduced even further. The recognition of continuous speech acts differs from the recognition of single words. In the latter case, gaps of several tenths of a second have to be left between every word which greatly reduces the difficulty of the task.

The recognition of continuous speech acts requires a signal to be broken down in order to identify the different words or phonemes within it. There are two different approaches.

The first consists of recreating a sentence from the signal. This approach is becoming increasingly popular. A signal is simply read without trying to understand its content; the focus lies entirely on decoding it. However, for every part of the sentence this method needs to search all of the words its dictionary contains. This operation is often of an unpredictable length of time, or simply takes too long.

Another approach uses anticipation to recognize words. This concept is similar to our brain which also uses the context to anticipate the end of a sentence. Equipped with "certain intelligence", the system can anticipate what will be said: a telephone conversation usually starts with "hello" and is followed by "Mr" or "Mrs", often followed by one of the names in the contacts list which the machine will then search for. This approach avoids the lengthy search in the dictionary and saves a lot of time.

In practice, no system is entirely based on a top-down or bottom-up approach only. In both cases, the vocabulary would be extremely limited. Operational systems use an approach that combines both technologies.

5.4.2. *Gesture recognition*

The use of gesture recognition as a human/machine interface used to transmit orders is as old as the first machines themselves. It is a hand or foot movement that transmits the order to the machine. Apart from the keyboard, the gesture recognition of computers also includes pens, graphics tablets, the mouse and other joysticks used for the first graphical applications. The keyboard as the sole interface no longer fulfils the task.

Today, these different types of interfaces are being improved by other sensors, for example, gloves which allow the machine to detect and interpret all forms of a body's gestures including facial expressions. However, an increasing amount of applications no longer need restraining receptors but use the signal of a camera (visible light or infrared) to detect movement. These cameras are today mainly used to follow hand movements as well as eye movements. Applications in this domain do not rely on conventional equipment but use algorithms to allow for an efficient interpretation of visual images by the computer.

The majority of the applications in this domain are used in experiments in virtual reality with in the field of sports. They are also used as simulations and moreover as a supporting tool that helps disabled people to communicate.

At the end of the 1990s, a group of researchers from the University of Carnegie Mellon, USA and the University of Karlsruhe, Germany combined vocal recognition and gesture recognition in a single interface. The aim was to create an interface which was capable of almost natural communication. It can read lips, interpret handwriting and even detect facial expressions and someone's eye movement.

5.4.3. *Processing and recognizing writing*

Processing and recognizing writing are specialized domains in the field of gesture recognition. This principle is already fairly well known and widely used in applications such as PDAs, smartphones and Tablet PCs which use a program that recognizes handwriting and touchscreen panels.

Handwriting recognition is an application that transforms handwritten texts into typed text. The application enables the user to write text with a special pen onto a window which is visible on the touchscreen panel and to obtain its transcription in the form of typed characters.

The new generation of mobile/portable Tablet PCs has this function which enables the user to take notes and use them in traditional or innovative ways.

Tablet PCs digitalize handwriting and transform it into typed characters with the help of an electromagnetic input/output device or a touchscreen panel. The handwriting is then analyzed by a recognition program which is part of Microsoft's Windows XP Edition Tablet PC. This program is based on several thousand handwritten pages in different languages which were used as a sample. The process of handwriting recognition identifies every individual word. It separates all lines and curls the user draws to write a word. By using a specialized pen the user can, for example, directly write his/her text on the screen of their Tablet PC. By drawing a box around a part of the text and ticking the center of it, this part of the text can be moved very easily as its movement is linked to that of the pen. This concept is called digital ink.

Two different types of devices are currently available: graphics tablets and touchscreens, i.e. screens that can be used as a digital notepad. Technical progress in devices that process either graphic images or handwriting is paving the way for new tools such as digital pens which could be used on any kind of material just like traditional pens. This kind of device is no longer dependant on the link between pens and digital notepads or pens and touchscreens.

> The displays of touchscreens are often made of glass and are equipped with piezoelectric transducers as emitters and receptors for the X (lines) axis and the Y (columns) axis. The touch screen's control unit sends an electrical signal of a high frequency (several MHz) to the transducer which then alters the ultrasonic waves in the glass. These waves are sent to the surface of the touchscreen with the help of a series of reflectors. These reflectors placed facing the transducers redirect these waves to the transducer which transforms them into an electrical signal.
>
> By touching the screen, the user absorbs part of the wave that goes through it. The received signal will be compared to the card, or the stored digital matrix, and the detected changes allow for the calculation of the coordinates of a specific point. This process works independently for the X (rows) and the Y (columns) axis. By calculating how much of the wave is absorbed, the Z axis, indicating depth, is also generated. The digital coordinates are transmitted to the computer for further processing.

A digital pen works just like a graphics tablet. It stores the trajectory $(x(t),y(t))$ of a pen and does not limit itself to the X and Y coordinates of an image, such as a bitmap. This trajectory enables us to understand the movement and potentially the recognition of handwriting, which is crucial information for software dealing with this issue. Intel's digital pen is a traditional pen that is equipped with miniaturized accelerometers and pressure sensors made up of silicon (the epitaxy of a layer of several microns of silicon in which strain gauges are used). The trajectory of every movement of the pen is captured, stored and transmitted to the computer which then converts these movements into a digital text. The paper (or any other surface that is used) apart from its traditional role of storing content, can be printed out with minuscule geometric images that indicate the positioning of the pen and allows for the identification of the page which is used in order to link it to a particular application.

IBM developed a digital clipboard which could replace traditional note pads. The digital clipboard captures the trajectory of a pen which is equipped with a transmitter using radio waves.

5.4.4. *Eye tracking*

With the era of the Internet, studies in the effectiveness of advertising messages that lead to the sales of commercial spaces, and research in the field of the visual aspects of pages that appear on screen have strongly developed over the past 10 years. This research has mainly led to the development of eye tracking. These systems have continually developed during this period. Over the years they have become more flexible and less restraining and "invasive" to the user. Today they pave the way for new applications in the human/machine interface.

Eye tracking is a method of capturing the position of our gaze, i.e. where we look. Amongst all techniques in this domain, the corneal reflection is one of the most commonly used. It allows for high precision of the data taken but does not restrain the user's movements. Several infrared light beams are sent by a group of diodes situated in the center of the pupil. The infrared reflections that are sent back by the cornea of the eye are detected and enable the user, after processing the data, to determine the center of the pupil as well as its position when focusing on the target object.

A system that analyzes the data can then retrace the eye movement of the pages that were looked at and gives an indication of which parts of the page were looked at for how long, the so-called GazeTrace.

The idea of using eye tracking as a means of communication and control for patients with physical disabilities led to the creation of a high number of research projects. 40,000 patients in Europe and the USA could possibly benefit from this kind of interface. However, currently only several hundred patients use this new form of technology. On the other hand, this type of application is far from being the only target for the technology of eye tracking. Any other user could be interested in this technology when wanting to use a system in situations where other parts of the body, such as legs and arms, are already used to carry out other tasks, i.e. while driving, operating an engine, etc.

Consulting and managing graphic environments in a virtual reality is based on other applications which are similar to this kind of interface. They are operated visually and are also very promising. The initial presentation of a page could be an (macroscopic or global) overview of graphics. The eye focuses on a section of the graphics and a zoom is used so this section increases its size in proportion to the rest of the page.

5.4.5. *Brain machine interface*

Brain machine interface (BMI) or brain computer interface (BCI) are interfaces that link the brain directly to the machine. This type of periphery is fundamentally different to all other kinds of interfaces between humans and machine: this kind of connection does not require any kind of transformation of a signal emitted by the brain's activity expressed in the activity of muscles (psychomotor). The latter is usually used as a signal transmitted to the machine. The speed of a usual chain reaction (brain, nerves, muscles, human/machine interface) can be now be decreased to several tenths of a second. Furthermore, other parts of the body such as hands, feet and eyes are free to carry out other tasks simultaneously. Military research is investigating this new technology which would enable a better coordination of fast and complex operations in a combat situation. Civilian applications may also take place in many different fields such as planes or using complex engines to carry out other tasks while driving, or in the field of crisis management, fires, etc. One possible application of this interface is to allow people who have lost their physical abilities or their senses to recover them artificially. In co-operation with the latest developments in the field of electronics, programming and cognitive science, a paralyzed person could drive a wheelchair with their thoughts.

Two approaches to this technology are currently being exploited.

The first approach uses the electromagnetic fields which are generated on a macroscopic level by virtual dipoles which represent the sum of all activities of a large number of neurons. In other words, all human activities generate a cerebral activity which is represented by the activation of the equivalent zones in the brain. When one of those zones is activated, a small electric current and a small magnetic field induced by the activity of 10^5 to 10^6 neurons can be measured. Two types of instruments allow them to be captured: electroencephalogram (EEG) or the magnetoencephalogram (MEG). The EEG measures the electric potential, i.e. the dipolar parallel feedback at the axis of the dipole. This feedback is diffused and affected by the tissue and sensitive to deeper lying sources. The MEG measures a magnetic field, i.e. a dipolar signal perpendicular to the dipole's axis. This type of feedback is very focused and hardly affected by the cerebral tissue. It is not very sensitive to deeper lying sources. Due to price and practicability, BCI interfaces are used in combination with EEG instruments.

Elements of Technology for Information Systems 175

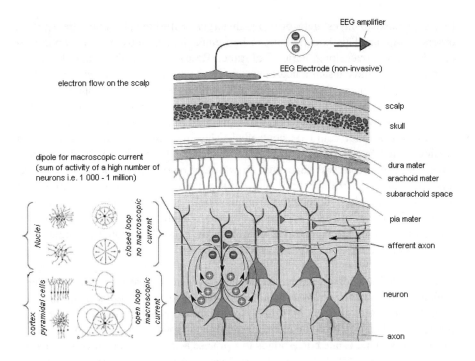

Figure 5.14. *The EEG, a macroscopic approach of non-invasive neural interfaces*

The EEG measures the macroscopic current underneath the scalp. Its simplicity represents a fundamental advantage in the development of cerebral interfaces: it is non-invasive, which means that it does not require an operation to implant electrodes or other linking devices to connect the brain to the system. The link is simply made through sensors which are placed on the head and are held in place by a hat or a hairband. These sensors capture cerebral waves and after decoding and processing them they transmit them to a computer.

The second approach is invasive and consists of implanting inorganic components which are connected to the neural system, i.e. directly implanted into the brain to create a link between movements and electric impulses. This approach is nearly microscopic and allows for the detection of the activity of a small group of specific neurons. However, this technology, as it is invasive and less reversible, is less adapted to operational cerebral interfaces which are ethically acceptable. It is used as a tool that allows for brain mapping, i.e. a detailed comprehension of the functioning of the brain, or more precisely the neural code which comes into play to activate or deactivate different neurons during all human activities. Before humans can finally use brain/machine interfaces to control the technical devices in their daily life, they need to investigate on a much larger scale how the brain transmits its orders to organs. This knowledge can be obtained by a study of data captured on neural activity over long periods of time. Researchers at the Duke University Medical Center have tested a neural system by capturing the signals at the level of a neuron with the help of several invasive electrodes which were implanted into the brain of an ape. The team of researchers was able to recombine the signals emitted on the neural level and, with the help of a program running on an external computer, the arms of a robot could be controlled. The ape was encouraged to move the artificial arms by providing it with food that could be reached with the artificial arms. The experiment was successful. It was later reproduced by sending the signal that controlled the artificial arms through the Internet and allowed the arms to be moved more than 1,000 kilometers from the ape's location. In 2002, the department of medical imagery at Duke University developed a telemetric circuit that allows the acquisition and the transmission of data through the skull without needing a sensor implant with cables which transmit information.

Since 2000, the results of research on non-invasive techniques which could be exploited industrially have breached all previously anticipated limits. The European project ABI (*Adaptive Brain Interfaces*) favors this approach and developed a cerebral interface based on the mental state of the individual in 2000/2001. In other words, every mental state corresponds to a precise activity. Equipped with an EEG on the person's head that measures cerebral waves, the user can learn to use the interface by forcing him/herself to present a clear and distinct mental state to the device each time s/he wishes to control the computer by his/her thoughts. For example: relaxation, mental focusing, imagination of moving the arm, a hand, etc. The system is based on a neural network that learns to associate the user's mental state to the respective activity. Five mental states are required to move the cursor on the screen. These are up, down, left, right and click. The ABI project showed that the user can play a computer game by simply thinking and edit a text with the help of a mental keyboard that appears on screen. Projects creating neural prostheses by BrainLab at the University of Georgia, USA use similar procedures. A non-invasive device captures the cerebral waves via EEG and enables the user to move a cursor by thought, or even command a wheelchair. At the beginning of the 1990s, the US Air Force were already carrying out experiments using BCI by enabling the user to control the flight simulator by thought alone. Only four non-invasive electrodes were used.

However, the researchers encountered two major difficulties. The first was the complexity of learning how to use the simulator. The users needed to concentrate in order to be in a mental state that enabled them to carry out a specific action. Furthermore, in order to interact with the machine and for it to run smoothly this mental state needed to be held and repeated several times. It takes several weeks or months for the user to learn how to interact with the machine effectively. The rhythm of choosing letters in the ABI project is very slow. 22 seconds are required per letter. This is far from BCI's goal to create an extremely high reactivity. The second difficulty is due to the procedures of a macroscopic nature. The required signal gets lost in a magna of "noise" or other undesired signals, and information is detected which is not useful. EEG waves are diffused and highly affected by the tissue; they represent the activity of nearly one million neurons. As the signal theory shows us, the risks of not detecting signals or detecting a false alert will be high until a reliable model that overcomes this weakness is found.

The usage of P300 waves currently represents a perspective in the process of creating EEG-based, non-invasive and more reliable cerebral interfaces. For about 30 years, neurobiologists have been observing reactions to exceptional but expected stimulus such as going through a red light or a stock market crash. After 400 milliseconds an electric signal which is typical for our brain is sent out, which neurologists call P300. The P300 wave indicates the cognitive process of decision-making. In 1999, a team of researchers at the University of Rochester, USA, showed that the P300 wave is the universal "symbol" of a thought linked to an "expected surprise" and always leads to a reaction of the subject itself and might involve the tools used (computer, machine vehicle, etc.). This universal reaction could be an essential property usable in systems of telepathic control.

> A team of researchers at the Technical University of Graz, Austria has developed a system that enabled a young paralyzed person to move objects with his left hand. This system combines the interface brain/machine (BCI) with an electric muscle stimulator. Thinking is an activity that requires our nerves and produces low electric currents within our brain. These currents are captured by the EEG, amplified and analyzed by a computer which selects those thoughts that express the desire to move the object from all of the patient's thoughts. This filtering process is possible because of one physiological particularity of our brain. Half a second before carrying out a movement, the brain produces a specific EEG that predicts the movement. By concentrating on that movement, the handicapped person, whose brain is linked to the BCI via electrodes, provides the computer with the required information which activates the muscle stimulator that is linked to the young man's forearm. Today, this young man can open and close his hand again simply thanks to the transmission of his thoughts.

5.4.6. *Electronic paper*

Trying to replace traditional ink and paper with electronic alternatives while preserving the advantages of traditionally used paper is something all electronic laboratories have been working on over the last decade. Electronic paper is a new visualization device whose qualities are flexibility and manipulation (i.e. it is flexible and fine), just like traditional paper but which can be updated electronically. It works due to electronic ink that is integrated in this device. Generating and updating the text consumes small amounts of energy. Maintaining the text on the paper so it can be read does not consume any energy.

This principle dates back to 1977. Nicholas Sheridon, a researcher at Xerox's Palo Alto Research Center (PARC), had the idea of replacing traditional ink by minuscule two-colored spheres placed between two slides of plastic. After 18 months of research the project was stopped as Xerox invented the laser printer. The emergence of the first generation electronic books and promising developments at MIT's Media Laboratory of applying electronic ink to paper, plastic and metal have encouraged Xerox to recommence research and development of Sheridon's idea.

The prototype of the electronic paper is called Gyricon (from Greek, *gyros*: turn around and imagine). Its base is a slide of flexible plastic on which a layer of electronic ink is applied. This ink consists of thousands of tiny bi-chromic pearls (one hemisphere is white, the other black) of a maximum size of 0.1 mm; they are covered by a transparent silicon film. Every hemisphere is of an opposed charge, these micro capsules are dipoles and able to turn around in reaction to an electric field. They are placed in a viscous liquid that allows for easier rotations according to the magnetic field which they are subject to and according to which they turn around. Letters or drawings are formed through the alternation of black and white points depending on which hemisphere the user can see. Gyricon's properties are very similar to paper or cardboard. It is less than 0.3 mm thick, which is very slim when compared to the 2 mm of traditional screens. Furthermore, Gyricon is flexible and can be rolled in a cylindrical form of 4 mm without any reduction in quality. It can also be folded at least 20 times before there are visible damages.

The micro capsules are produces by applying melted wax on two opposing sides of a rotating disk. The centrifugal force moves the wax towards the outside of the disk where it forms black and white filaments on the respective sides and then two-colored spheres which become solid very quickly.

The micro capsules are applied to a silicon film and all of it is covered by an oil of low viscosity. All of this is finally heated. The heat leads to the expansion of the silicon film and oil can enter the gaps between the micro capsules. At the end of the operation every micro capsule floats in its own oil bubble which allows them to turn around easily and respond to an electric field. The required time to turn is 80 to 100 ms. This time could be shortened even further by decreasing the diameter of the micro capsules.

Elements of Technology for Information Systems 179

The silicon layer containing the micro capsules is placed between two slides of plastic or glass that contain a network of electrodes that is very similar to that used for screens made up of liquid crystals. These electrodes direct the rotation movement of the micro capsules. Either the black or the white side of these spheres will be visible and every single one of them reacts immediately and independently of the neighboring micro capsules. When using plastic for this device the final sheet of paper also becomes flexible.

Reading a text written on *e-paper* requires ambient light just as is the case for traditional paper. Energy is only needed to change the given information. This device uses 2% of the energy consumed by the first generation of electronic books which relied on LCD screens.

In December 2000, Xerox announced the creation of Gyricon Media Inc. which is in charge of the commercialization of Xerox's electronic paper. This company will develop applications for visual identification and for putting up posters in stores.

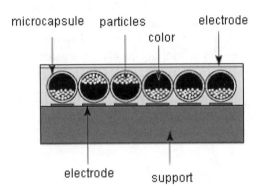

Figure 5.15. *Make up of a layer of electronic ink: white micro pearls in a capsule filled with liquid black color*

MIT's Media Laboratory has used Nicholas Sheridon's idea since the 1990s and has been developing a different variety of electronic paper. They used white micro pearls in a capsule filled with liquid black color. The micro pearls used are spheres of 30 to 40 microns in a transparent shell. They contain a titanium dioxide which is white and shiny. When exposed to an electric field these micro pearls move to the surface of the device. This slightly different procedure simplifies the fabrication of spheres and improves the final resolution. MIT's researchers founded E-ink in 1997 to commercialize their product and produce it industrially. Their method allowed for the introduction of color by using colored pigments in the spheres. Researchers from

Bell laboratories had the idea of using bacteriorhodopsin (see section 3.6.3) when introducing color to electronic paper as bacteriorhodopsin changes its color when exposed to an electric field.

> The word *e-paper* (electronic paper) has different meanings. In Europe, it is often used for online editions of newspapers and magazines. In the USA, the term is used in its literal sense. Here an *e-paper* is a visualization device made up of a plastic film which is fine and flexible and similar to traditional paper but which allows for the information to be changed either by a wired connection or via a radio signal. A text is written on this kind of paper via electric impulses and not by using traditional ink.

However, these devices remain limited to black and white in the visualization process. Over the last decade, Philips' laboratories have been working on a principle called electro-wetting which allows for colors to be used in the visualization process.

Every point or pixel consists of a bubble that contains water and a drop of colored oil. By applying an electric field the oil is compressed and the white side of the bubble appears. When there is no electric field applied to it the oil spreads out and its color can be seen. By controlling the electric tension the intensity of the color can be adjusted.

This device would be a fairly normal visualization device if it was not capable of visualizing color and, moreover, making rapid changes to its images (e.g. the image can be changed in 10 ms, which enables the user to watch a video of 100 images per second).

However, this is not the only advantage this device has in comparison with traditionally used devices. This system offers an excellent legibility under all conditions, a very impressive luminosity and a very low energy consumption. As its construction is much simpler than a screen of liquid crystals, it uses less energy and sends out four times more light mainly because the color is obtained from the colored oil and not by colored filters which are used in flat screens, for example. Just as for traditional paper, the legibility is very good because natural light is used to create a contrast instead of a backlight. Not only are the contrasts much better but the traditional disadvantage of screens made up of liquid crystals is that they can only be looked at from a certain angle, however, this is no longer the case. The absence of the backlight is also an advantage when it comes to energy consumption. In contrast to traditional screens, no energy is required to maintain the image on screen. Only modifications such as changing the image consume energy.

Last but not least, the ductility of these new screens should allow for the development of less fragile devices which could be used everywhere. This fact paves the way for fixing screens to our clothes, a concept which is known as *wearable computing*.

5.4.7. *New visualization systems*

The general user often groups generations of computers according to their storage and their displays. Punched cards were used in the 1960s. In the 1970s the black and white cathod ray tubes (CRT) emerged and were used until the 1980s. The end of the 1980s is associated with color CRT screens. Multimedia systems appeared in the 1990s ranging from Liquid Crystal Display (LCD) to plasma screens (PDP or *Plasma Discharge Panels*). Today's computers are often seen in connection with the Internet. This shows to what extent visualization systems are part of the image of a computer.

Considering the future of computers, therefore, also always includes the future of visualization systems that reproduce data, and, even more so, developments in screens.

The beginning of the 21st century is marked by the replacement of cathode ray tubes with LCD screens or laptops moving on to systems of even smaller dimensions. The TV market has also benefited from this change. After 50 years of only using CRT, plasma screens and LCD displays have entered our homes. They are larger and used not only as TVs but also for watching films or other video documents. The TV is as a "multifunctional visualization tool", a multimedia screen which covers the main TV channels as well as video documents on magnetic disks such as DVDs and video games, or simply as a larger PC screen. In companies, large flat screens such as plasma screens or LCD beamers have become indispensable tools for group work that uses computers.

However, the successors of the CRT technology, be it LCD or PDP, will not have the same economic lifespan as their predecessors, even if their technical lifespan is nearly twice as long as that of CRT devices. New solutions might enter the market in the next five to seven years.

OLED (*Organic Light-Emitting Diode*) technology, which is mainly being developed by Kodak, is the next generation of flat screens adapted to color videos. Its performance in terms of sheen and a clear picture cannot be matched by other technologies currently available. In contrast to LCDs, OLED screens provide their own source of light and do not require a backlight. As a backlight is no longer needed screens are much slimmer, very compact and very light. Furthermore, OLED offers a large angle of up to 160 degrees from which the image can be seen even in full light. Last but not least, its energy consumption of 2 to 10 V shows its efficiency and also reduces the heat that electric devices produce due to electric interferences.

OLED cells are light emitting diodes (LED) whose light emitting material is organic. As for an LED, the material becomes fluorescent when an electric current travels through it. An OLED cell consists of several layers of organic material which are in between two conductors. The first conductor is a transparent anode coated with indium tin oxide (ITO). The second conductor is a metallic cathode. The first generation of OLEDs had only one layer of organic material between the two conductors. Their performance was very weak. Today's OLED cells consist of several layers: one injection layer, one transport layer for holes, one emitting layer and one transport layer for electrons. When the appropriate electrical current is applied to the cell, positive (holes p^+) and negative (electron e^-) charges are recombined in the emitting layer to produce electroluminescent light. The structure of the organic layers as well as the anode and the cathode are created in a certain way to optimize the recombination process in the emitting layer which also increases the emission of light from the OLED device.

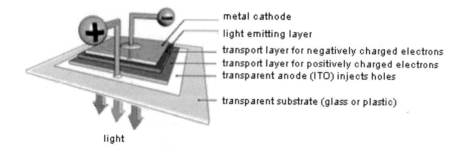

Figure 5.16. *The structure of an OLED cell*

OLED first appeared in 1997. Several companies have been interested in this technology ever since. Sanyo, Pioneer and Motorola are some of them. The global market for OLED screens has evolved from some hundred thousand units sold in 1999 to a turnover of $3 million in this field and several hundred million screens sold in 2005. In other words, the turnover was over $1 billion in 2005. The reason for this marked increase is mainly the strong demand for MP3 players, PDAs and digital phones of the new generation. All of these were the first systems to make OLED technology available to the general public, even if the size of the screens still remains rather small.

OLED is not the only possible successor of LCD screens; other possibilities are currently being researched. An example is FED (*Field Emission Display*) visualization which uses field effects. This technology is based on the amplification of an electric field produced by point effects which arise in the strong electric field around small-radius charged bodies or electrodes (see for example electric arcs, or why lighting hits the highest points).

The cells of an FED screen consist of two parts that are sealed in a vacuum: the cathode and the anode. The cathode disposes of several thousand micropoints per pixel placed in the holes of a size of one micron per diameter. When the grid of the conductor is placed near the point and a potential of several dozen volts is applied to it in relation to the cathodic conductor, the electric field is amplified at the points. This amplification leads to electrons being released by the field effect.

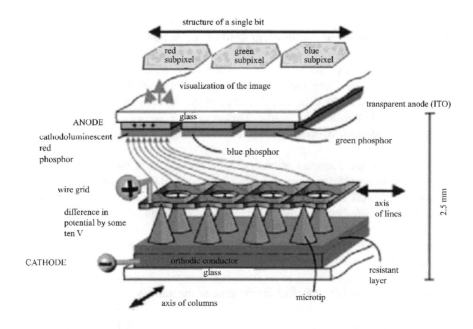

Figure 5.17. *FED (Field Emission Display)*

Due to the application of another electric field between the cathode and the anode, the released electrons are drawn towards the anode. The anode is covered in beads of cathodoluminescent phosphor which emit light when placed under an electron beam. The image can be seen from opposite the anode.

The development of FEDs will never use microtip technology. Due to their small size (1 to 2 mm), they cannot be produced on a large scale and can therefore not be exploited cheaply. Releasing electrons by carbon nanotubes which would allow the potential required to be decreased for emission could be a possible alternative when it comes to industrial applications. Motorola abandoned its research in the field of microtips in 2001 and is now investigating the electronic emission of nanotubes.

In 1986, the first monochromic screen-based on FED technology was presented by a team from LETI (Laboratory for Electronics and Information Technology). This team worked with an industrial partner called Pixtech to develop the production and the characteristics of FED screens. However, their research could not be exploited on a commercial level. Today, it is too early to say whether FED is a competitor to OLED screens. However, OLED screens will be used over the next 10 years and are LCD's successor in this field.

5.5. Telecommunications – a different kind of revolution

Until quite recently, the telephone was the only means of telecommunication between two people. From the 1970s onwards computers increasingly used this infrastructure to exchange data. The exchange of information requires a much higher speed of transmission than the simple transport of a voice in a telephone network. The capacities of our networks needed to be increased on a very broad scale.

The world of telecommunications changed drastically at the end of the 1990s when the circulation of data exceeded that of the voice. The operators reformed their infrastructures to offer their clients ever more integrated services. The development of computers is therefore strongly linked to developments in telecommunications.

New infrastructures use digital networks and rely on the circulation of packets of information. Until the 1990s, telecommunication networks worked in analog mode. In other words, if a car wanted to go from city A to city B, the entire road between the two cities needed to be blocked for the time of the journey.

On the other hand, the circulation of packets of information allows several vehicles to use the street at the same time. The deregulation of the telecommunications market, and more particularly of the local loop (also referred to as a subscriber line) has largely accelerated the development of networks which offer several services at the same time.

The general user could estimate the progress made in the field of computers while developments in the field of telecommunications were unpredictable. However, with telecommunications via satellite, the improvement in laser technology and the progress in the field of fiber optics, which has rapidly taken over from copper, the telecommunications industry is blossoming.

This trend will lead to important changes in telecommunications once a global movement of deregulation sets in and paves the way for competition on the market and the increasing popularity of the Internet. The Internet with its Internet Protocol (IP) can simultaneously diffuse all forms of communication. A packet of information can carry voice as well as image. Web solutions will unleash the imagination of providers in terms of content as well as service and will also change consumers' habits.

From its invention onwards, the price of telecommunications remained rather stable. However, there has been a significant drop in the price over the last 10 years. All fields of telecommunications such as voice, data or video have become very cheap. Even video conferences, which were previously reserved to very large companies, have now made their way into general households.

The next step in this revolution which will favor the development of ambient intelligence and diffuse informatics is based on wireless connections. This new technology and the protocols that are used are playing a crucial role in the development of emerging platforms (see section 4.3.2) and third generation (3G) mobile telephones. These cell phones allow for an extremely high speed of information exchange. Telecommunications play a central role in the development of web diffusion.

The future of telecommunications lies in its network which could consist simply of nodes in the form of terminals. These nodes are the terminals themselves (cell phones, PDAs, laptops, etc.) which manage themselves and could replace stable antennae.

This analog principle called P2P (see section 4.3.3) is perfectly adapted to the introduction of a highly interlinked network in which the objects that communicate become omnipresent. The basic idea is the same as the one that gave rise to the ARPANET[6] network. What would happen if the antennae were damaged by a natural disaster?

---

6 In the 1960s, the United States Department of Defense (DoD) commissioned a study on the reliability of communication networks during armed conflict. The famous and worrying report was given by the RAND Corporation, which is why this document is known as the "Rand Report". The report stated that the communication infrastructure is extremely centralized and lacks autonomy of nodes which could take over in case a connection is cut between some of those points. If a secondary structure failed to provide its service, it could endanger the entire system. In the case of even a small nuclear attack the entire telecommunications network of the USA would be at risk. This report led to the creation of the ARPANET network, the

This danger could be avoided if mobile terminals serve as antennae within the network. The link is therefore made directly between two terminals that would like to communicate with each other. The solution also seems very convincing because stable antennae that cause permanent magnetic smog, and are therefore a danger to the public, would no longer be needed.

Even if this system seems very convincing, due to the absence of a central unit (i.e. swarm intelligence; see section 6.4.2), several problems are still linked to it. These are, for example, the confidentiality of the transmitted information, its quality, as well as the speed at which information can be transferred.

It will still take three to five years for these networks to be made available. However, a good knowledge of the density of these nodes is currently being established.

> A system that manages itself can only be created if security can be guaranteed. In current networks, the protection of privacy and personal data protection are directly ensured by the provider. A self-organized network would no longer have such central control.
>
> The nodes would directly exchange their digital security codes and their addresses. In order to do so, the nodes, or rather terminals[7], need to be connected directly at least once to exchange their digital security codes and their addresses. This is the coupling process. This direct link would mean that the users of the terminals "know each other". Terminals A and B "know each other" and can establish a secure connection. Their communication cannot be intercepted. If B knows the digital security code and the address of C, A can contact C via B without C knowing all security codes.
>
> For a long time people believed that the danger for the security of independent networks lay in weak links within it. Simulations, however, have shown the opposite. This type of peer-to-peer relation could actually increase security.

---

predecessor to the Internet that is based on the following principles: tested and disposable equipment, shared connections, costs shared between the operators and field extensions.

7 In the short term, devices (e.g. objects or web services which indicate their properties and allow for them to be used) will show their methods online and allow other devices to check and manipulate them.

5.6. The triumph of microsystems

MEMS are made up of components between 10 to 100 microns in size (i.e. 0.1 to 0.01 mm). They usually consist of a central unit that processes data, the microprocessor and several components that interact with the outside such as microsensors for pressure, temperature, chemical or magnetic changes as well as one or more microactuators. Microsysems need to fulfill highly contradictory standards. They need to offer best performance at the lowest cost and with the least overload of data.

The first application that could be could be considered as a MEMS was a field effect transistor with a wired grid that was created by Westinghouse in 1969, a product that turned out to be more of an object of curiosity than something that could be exploited industrially. However, it still marks the beginning of this technology. At the beginning of the 1970s, the producers used engraved substrate plates to produce pressure sensors. At the beginning of the 1980s, research had moved on and the technique of micro-manufacturing was used to create actuators made up of silicon polycrystalline used in the read heads of disks. At the end of the 1980s, the potential of MEMS was widely recognized and their applications increasingly entered the fields of microelectronics and biomedicine.

In the 1990s MEMS received special attention, especially from the USA and government agencies offered support for projects in this field. During this period, the Air Force Office of Scientific Research (AFOSR) strengthened their research into materials and DARPA created its foundry service in 1993. At the same time, the National Institute of Standards and Technology (NIST) offered support to private foundries which were working in the fields of MEMS or produced CMOS components.

At the end of the 1990s the MEMS device was globally produced industrially. Factories that produced MEMS were built by companies such as Bosch and Motorola. The US government is still very interested in MEMS and continues to subsidize agencies such as DARPA. Recently, a series of miniaturized systems were used in order to control the performances of a space shuttle sent out by NASA. These systems were used for navigation, detection, propulsion, digitalization of data as well as to control temperature.

These systems that emerged more than 30 years ago have been omnipresent in our lives since the 1990s. They can be found in ABS systems, computer applications such as ink jet printing cartridges, all every day multimedia applications, cardiac pacemakers, and in the field of biology where lab-on-a-chip applications allow tests to be carried out, inside the target organism which avoids high costs.

In the 21st century, microsystems are mostly used in cars. A luxury car in 2003 contained as many electronics as an Airbus A310 did in the 1980s. In 1974, 4% of a car's price represented the electronics used in it while in 2003 this figure increased up to 20%. In 2010 the estimated percentage will reach 35%.

> Microsystems in the car industry have developed a great deal from their starting point, which was the "intelligent" pressure sensor for tires. MEMS are now a vital part of car systems and fulfill all different kinds of functions such as the accelerometer for airbags, sensor for petrol, controlling the breaking process and, last but not least, reducing the noise in the passenger compartment. Part of the latest innovation is "intelligent" tires that indicate to the driver if one of the tires is going flat. Their performance is so sophisticated that they are able to provide this information up to 80 kilometers before the tire eventually needs to be replaced.

Many other sectors also need to be mentioned here. Especially in the field of telecommunications, microsystems are used on very widespread hyperfrequencies. In the field of cell phones and satellite connections, optical microsystems are used for the optoelectronic conversion.

The fields of aeronautics, space travel and military, which first used microsystems, are still developing new projects. There is no need to give a list of all applications. Let us simply mention the micropyrotechnical systems which pave the way for applications such as microsatellites which weigh about 20 kg. Another interesting example is the research project that is trying to replace the flaps on airplane wings with a network of microflaps (similar to the feathers of a bird) in order to increase the performance in terms of security and fuel consumption.

Microsystems are currently undergoing strong industrial development and represent a market of €50 billion. Their growth rate lies at 15% per year for the last 10 years. This is a market that will grow even more as the production of microsystems will multiply tenfold in the next decade.

Elements of Technology for Information Systems   189

Microsystems growth forecasts

Biometric sensors market CAGR 1999-2005	40%
Lab-on-chip (LOC) sales CAGR 2002-2008	33%
LOC detection of biological/chemical agents market CAGR 1999-2004	20%
Microfluidic device sales annual growth 2002-2008	30%
Biochips US market annual growth 2002-2006	23%
Biomedical microsystems global market annual growth 2001-2011	20%
Medical applications using semi-conductors market annual growth 2004-2009	18%
Fiber optic component assembly/test products sales annual growth 2000-2005	21%

Source: Canadian Microelectronics Corporation, annual report 2004

Table 5.1. *Outlook on the growth of microsystems*

There are two main reasons that explain their commercial success and positive predictions for their future development. First of all, MEMS have shown reliable performance under extreme conditions and within extremely hostile environments (e.g. thermal shocks, extreme pressure, humid or salty environments, etc.). This fact makes MEMS an ideal candidate for all possible applications in different fields. The second reason for their success is that MEMS can be used for the mass market (e.g. car industry, telecommunications, household appliances, audio and video application, etc.). Therefore, their production is financially beneficial as is the creation of new applications.

The real challenge of these devices lies, however, in the rise of interconnected microsystems that are able to interact with one another, just like a colony of tiny intelligent objects. A new era of information systems has begun. In the past, the data processing units were centralized just like the brain of a primate. Now, the distribution of intelligence will move through the entire network. The diffusion of this decision-making power paves the way for four brilliant principles. These are increased calculating capacities, reactivity towards local situations, a tolerance towards problems (see section 5.5) and self-adaptation.

> In 1991, Goss and Deneubourg showed in a simulation how a group of simple robots, which could represent a colony of microsystems working in a network, manage to navigate in an unknown space without a map, visual recognition, or the slightest point of orientation. The robots perceive the given space as smaller than it actually is and can encounter obstacles. However, these robots are able to collect objects which are scattered on a square surface and move them to the center of the space.
>
> Even if the method of using colonies is currently less efficient than complex centralized systems, it beats all other competitors when it comes to simplicity, reliability and flexibility (adaptability).

5.7. Is this the end of the silicon era?

The question of whether the domain of silicon will finally reach its limits in the near future (i.e. 5 to 10 years) is far from being answered. This book does not aim to enter the Manichean debate of experts but would like to show that this domain could be combined with, and even benefit from, new trends in technology. More precisely, how can the value chain reorganize itself, open up opportunities and break the deadlocks that go hand in hand with these opportunities?

> The debate on whether silicon will be a pillar of nanotechnology is similar to discussing what came first, the egg or the chicken. The term nanotechnology usually refers to processes, structures, materials and devices which are smaller than 100 nanometers. This term is used no matter whether traditional top-down processes or bottom-up assemblies are referred to. The controversy on this issue is brought about by the "purists" who base their ideas on a neologism under which only the second approach i.e. self-assembly and structures on an atomic scale are acceptable.
>
> There is, however, a limit of quantum phenomena which will sooner or later come into force when traditional processes are used to work on a scale less than 50 nm. Since 1900, and mainly thanks to the research carried out by Max Planck, it has been known that energetic exchanges i.e. heat, electric current, light, etc. can only be carried out in small packets of elementary particles, the quanta, when working on a nanoscopic level.
>
> Even if quantum phenomena are barely controlled and understood in the scientific and industrial sector, they still have a positive impact on certain devices that are currently used.

Elements of Technology for Information Systems   191

Since 2006, major companies that dominate, and therefore often limit the market of traditional microelectronics (Intel, Texas Instruments, AMD, etc.) have started working on the engraving process at 45 nm. This means that there are more than 20 million transistors positioned per meter. Only seven years ago experts considered such a high level of integration impossible.

At this level of miniaturization the devices are comparable to the size of a virus, they are therefore smaller than living organisms. Some layers of the material used are even slimmer than a virus. The new generation of transistors made from silicon oxide is only five molecules thick. We have, therefore, finally entered the domain of nanotechnologies. In other words, a domain where measuring units are on an atomic scale.

Over the course of 30 years the semi-conductor industry has greatly evolved. The amount of money spent in R&D facilities is higher than ever before and exceeds the money invested in the Apollo program that led to the landing on the moon. This figure currently makes up 15% of the turnover in the sector which is between €22 and €30 billion per year. In less than 15 years the three biggest European producers have multiplied their investment tenfold and have forced out an increasing number of players in this industry (see section 2.3.1).

During this period, this domain and the computer industry have profoundly reformed their value chain. What is usually called the "golden age of information technology" and marked by the arrival of the PC on the market also marks the beginning of a transition in organization. The trend moves from a vertical integration which is dominated by a handful of multinational players towards an economic distribution chain marked by a large network of companies which co-operate with one another. One form of co-operation is based on a traditional hierarchical relationship of sub-contractors who then work together as a form of strategic co-operation such as partnerships, consortiums, joint-ventures and other agreements made in technical sales. Everybody does business with everyone else, but even a commercial co-operation does not mean that those companies get along well.

> After the new trend was introduced by the arrival of the PC, the emergence of open systems such as UNIX and the first standards that were set in terms of creating open systems has led to all partners being focused on a shared objective.

The reorganization process of medium and large size companies was perceived as progress in information systems by clients at the time. Vertical integration enabled technical directors and managing directors to relax. Companies such as IBM, DEC, Siemens, ICL and Bull were in charge of all technologies, produced all devices from processors to programs and even added maintenance devices to their systems. The price was rather stable since there was not much competition on the market.

The new value chain that is currently emerging around open systems is questionable in terms of global responsibility as well as in terms of quality. On this note, the new value chain cannot compete with the previous system that was dominated by large companies. However, this new trend has opened up new segments in the market. These are the creation of small companies and a strong increase in usage. Information technology has entered the period of democratization. Parts of the economy might eventually by recovered by bigger companies which would eventually lead to a change in the criteria of profitability of the investment made in the field of information systems. Furthermore, this open value chain based on a virtual network of companies is extremely flexible and has adapted itself perfectly to the fluctuations of a market which has become more and more difficult to enter.

> Between 1999 and 2000, the benefits of virtual value chains and of companies that co-operate were mainly seen as a product of the Internet.
>
> These approaches require a system of organization as well as a communication tool that allows for co-operation. The Internet was therefore the ideal channel to be used. Otherwise, it would not make sense to outsource part of a company if the delivery of a quality product according to the deadline can no longer be guaranteed.
>
> In fact, the concept of meta-companies had been introduced by much older systems of information exchange. In the 1950s, they were discreetly introduced in the car industry and in aeronautics. They were based on exchanges of documents printed in a structured manner. In the 1980s, the progress of viewdata and the exchange of data called EDI and EDIFACT standards further improved the situation.
>
> Western car producers realized that Japanese companies were making great progress. Before it was too late they understood the secret which lay in the rapid and efficient communication with the sub-contractors and could therefore take action. The ODETTE organization that worked on the electronic exchange of data in the European car industry from the mid-1980s onwards can be considered as the first European meta-company.

Will the next step in technology erase the limits of information technology that are set by quantum phenomena? Will this be the beginning of electronics in molecular dimensions? Maybe the next step will not erase those limits straight away.

Elements of Technology for Information Systems   193

A report which was established by the Rand Corporation has investigated the interlinking of three key areas of technological research and development by 2015. These key areas are nanotechnologies, new materials and biology.

Taking into account the unification process of these scientific disciplines and the increasing transversality between them, the report showed that microsystems will play an important role. They use a traditional form of technology but distribute the computing power throughout the entire network. The technology basically remains the same but new applications will lead to visible changes.

At the same time, the first devices using nanotechnologies which are in extremely early stages of research will progressively be introduced into different applications in the next 15 years.

In other words, the development that is perceived by the user will be shown once more by the nature of its applications (i.e. dilution of microsystems) rather than by the technology itself. Nanotechnologies will go hand in hand with the trend towards spreading calculating power throughout the entire network which will mark the "new information technologies".

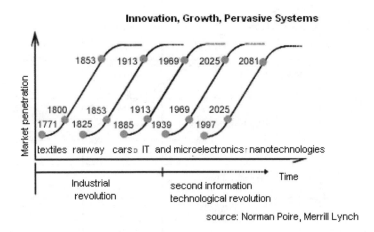

Figure 5.18. *Innovation, growth and level of pervasive systems, periodic cycles*

The end of the silicon cycle does not symbolize a rupture. Rather, there will be a gradually increasing crossover between technologies similar to the phenomenon of the development from vacuum tubes to semi-conductors in the 1970s, or the development from the logical wired machines towards a digital technology that was micro-coded in the 1980s.

194   Nanocomputers and Swarm Intelligence

During this period, the user observed the change from black and white to color TV and discovered other digital devices such as teletext. However, the question of whether a TV is based on vacuum tubes, transistors or is equipped with a microprocessor leaves the user rather indifferent.

A realistic outlook shows that silicon will dominate the market for another 12 years. This decade will allow applications using nanotechnologies to emerge on an industrial scale and will leave time to resolve problems such as the assembly of a large number of very small objects.

Until then, the semi-conductor industry will rely on an extremely competitive production system in terms of cost and mass production of the components. They own the monopoly of an industrial process that has reached its final stage.

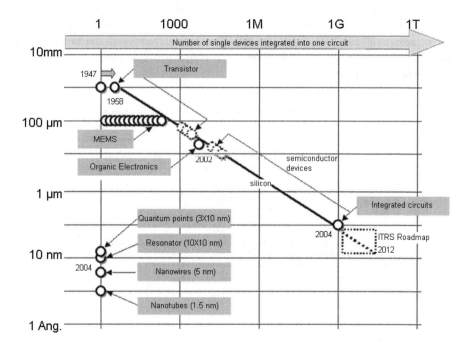

Figure 5.19. *The assembly of a large number of very small devices can only be carried out in the field of silicon. In the short term, building plans for computers and the infrastructure of information systems do not have an alternative to the domain of traditional semi-conductors (source STM)*

The beginning of the nanotechnology industrial cycle will concentrate on high performance components while the silicon industry will continue to produce peripheral components. The production process will run at an optimal level and lead to an extremely high cost efficiency.

These changes in technology do not necessarily lead to changes in the players involved. The silicon domain is one of the industries that has been most commercialized throughout history. However, it also remains an example that can be applied to other fields. Because of their financial resources in the field of research and development, the companies which are focusing on silicon right now might be the first ones to propose industrial applications for nanotechnologies. IBM or Intel, amongst other companies previously mentioned, have shown that they are the forerunners in this field. Before entering the era of the quantum computer there are still many different aspects to be researched and exploited. In this way, it will be possible to push the limits set out by Moore's Law even further and also reduce the cost for the production of a single device.

> The real challenge for nanotechnologies compared to traditional photolithography on silicon lies in the aim to industrially produce logic gates of 20 nm or smaller in the next 12 to 15 years. The integration should be at more than 10 billion transistors per chip, which can be compared to assembling 10 billion bristles on a synthetic soccer pitch. Miniaturization also leads to an increased speed in operations. Several hundred thousand electrons were previously necessary to change one logic gate from "0" to "1". Nanotechnologies only use one single electron or only a few electrons to carry out this change. With an increased speed and a reduced energy consumption, the work cycle of those new generations of machines will be more than 100 GHz as opposed to today's 3 to 4 GHz. As the work cycle is directly linked to the heat that is produced during operations, the transfer of the charge of hundreds of billions of electrons per operation would melt a traditional silicon chip. The work cycle could not be higher than several GHz, at least for industrial applications.

Chapter 6

Business Mutation and Digital Opportunities in the 21st Century

6.1. Towards a new concept of information technology

The definition of a computer has undergone a significant transformation process. The predecessors of the currently used machines were all produced for the specific task they would carry out. Some examples include: Chinese abacus, Pascal's calculator, Charles Babbage's difference engine, Hollerith's mechanical tabulator, analog guidance computers in A4 and V2 rockets, digital guidance computer for the Apollo missions with astronauts, etc.

In the 1980s, analog computers were given preference over digital applications when used during long-term simulations.

Differential equations that needed to be solved were too complex for the digital computers that were used at the time (physics, mechanics, nuclear applications).

The ever cheaper development and production of microprocessors will set an end to analog computers before the end of this decade. Given its ever increasing performance and capacity to solve arithmetical or logical problems, the PC, as the universal tool for processing information, will take over everything else.

> On 12 August 1981, IBM launched the PC (*Personal Computer*), which was a universal 16 bit machine using an Intel 8088 processor of 8/16 bits with a memory of 16 kBytes. The PC cost $1,665.

During the last decade the first global networks (e.g. Arpanet) and high-speed local networks (e.g. Ethernet) paved the way for the distribution of data.

In the 1990s, a second generation of information technology emerged with the rise of the Internet and web applications that were based on the concept of hypertext. "Everything free for the free"[1] stated Nietzsche in his famous aphorism. Tim Berner-Lee's ideas which lead to the creation of the World Wide Web in beginning of the 1990s were largely based on this Nietzsche's concept. "From the moment that information becomes available it needs to be made accessible to everybody."

Access to knowledge was changed just as much as it had in the case of Gutenberg's invention. Being computer literate is a new criterion of literacy just as being able to read and write was a criterion from the 15th century onwards when printed documents and books entered the era of democratization.

> Traditionally, a text is written and read "bit by bit" in sequences. This tradition stems from the material that was used to produce written documents and remains the same for a roll of papyrus in antiquity as well as for today's books.
>
> Hypertext is a non-linear presentation of a text. It remains a text but the reader is no longer obliged to read it from the beginning to the end. From information to entertainment, all are presented in hypertext. A text is read while the navigation process takes place in hypertext.
>
> For a long time, books have been written without the aim being for them to be read from beginning to end. Some examples are dictionaries or certain specialized books which are used for consultation only. This kind of book was the first to have great success when it was transcribed using hypertext.
>
> Today it remains rather difficult to imagine a similar process in the genre of classical literature, for example, a novel with a clear storyline.
>
> However, a large number of modern novels no longer use linear storylines. For about a century, even very high-ranking literature has not followed linear storylines. This is why a hypertext can be created for it. However, this domain is still in the experimental phase.
>
> In the media, the trend goes towards visual effects such as images, videos, simulations or animations. Hypertext can therefore now be referred to as hypermedia. Hypermedia is, however, very different from multimedia, a term which is currently very widely used. Multimedia stands for the possibility of using different forms of media at the same time. Hypermedia, however, refers to several types of media which can be navigated through hypertext.

---

1 Original quote: "...dass, den freien alles Frei stehe".

The developments in information technology have led to the emergence of highly interconnected systems of the current generation[2]. The next generation of computers will be integrated into the surrounding area it is controlling. In order to do so, intense miniaturization is required. Furthermore, these devices need to communicate with their environment or with other devices in micro networks.

As control units will be integrated into everyday objects, allowing for even easier access to information, a change in information technology will take place. The universal model of the PC will be replaced by more specialized systems which will be miniaturized. In a way, information technology is going back to its roots.

The integration of computers into their environment does not require technological platforms of the post-silicon era or other radical alternatives. The first steps towards an omnipresent digital world have already been taken. The computer, being the predecessor of the future's information technology, will therefore be transformed.

However, a simple increase in performance is not sufficient to lead to such a transformation process. For a technology to become part of our daily lives, as was the case with, for example, electricity, transport and medication (i.e. permeating technology), two criteria have to be met. The price needs to decrease steadily without curbing the technology's performance. Furthermore, the technology's performance needs to increase steadily.

6.2. Ubiquitous information technology and the concept of "diluted" information systems

Whether they are still part of the silicon era or use alternative solutions, the next generation of information systems will entirely change our idea of information technology. Developments in IT are mainly characterized by miniaturization and democratization. However, the ever increasing level of the device's integration into its environment also needs to be taken into account.

The next generation of information technology will no longer be limited to workstations, PCs and servers which offer a high capacity for processing and storing data, or the successor of currently used mass storage that can hold entire libraries on a hard disk the size of a matchbox. The new generation of information technology will go even further and these computers will interact in a network of a previously

---

2 With 450,000 servers in 25 data processing centers in the entire world, Google is able to process several billion search entries a day. In 2006 Microsoft used 200,000 servers for similar tasks and plans to dispose of 800,000 servers by 2011.

unthinkable bandwidth. The Internet will move from high-speed over very high-speed to hyper-speed. These developments will simply follow the axis of progress made in digital technology.

The ubiquitous revolution is not only changing the tools used but also the approach towards information technology.

> Ubiquity: describes the fact that micro-organisms can be found everywhere.
>
> The concept of Ambient Intelligence (AmI) is based on a network of computers that is integrated in the environment of our daily lives. Human beings are surrounded by intelligent interfaces that are integrated into familiar objects through which the access to a number of services is made possible in a simple, comfortable and intuitive way. The number of these familiar objects is increasing and the way they process and interact introduces the notion of an "augmented reality" (AR)[3].
>
> Ubiquitous information technology and ambient intelligence increase the amount of all different kinds of data that can be processed by using ever lighter and more specialized computers (e.g. PDAs, cell phones, chips for home automation, cars, health sector, etc.). By making access to data easier, ubiquitous information technology and ambient intelligence also introduce new risks for the confidentiality of data.
>
> In the 1980s, Xerox PARC developed the predecessor to these systems when working on media platforms and their properties. This concept was first called embodied virtuality. In 1991 Mark Weiser introduced the term ubiquitous computing to describe the phenomenon.

This "new age" of information technology can be found in the concept of communicating objects, intelligent devices, pervasive computing and web diffusion. This trend is leading us towards a world in which information technology will disappear because it is integrated everywhere. This trend is, to a certain extent, the counterpart of virtual reality.

Virtual reality makes human beings part of the computer while ubiquitous computing forces information technology to be adapted to, and embedded into, our natural environment.

This new concept of computers changes the notion of our desk in the office as the gateway to the world of information. The concept of the PC will be questioned.

---

3 Augmented reality has several definitions. The most commonly used definition describes information systems which allow the addition of a 3D or 2D model of the world that surrounds us. Augmented reality is used to evaluate architectural projects or to insert advertisements into videos that broadcast sporting events.

This will be an important change as the relation between human beings and computers has not changed over the last 25 years, even though very important changes in the field of IT have taken place during this period.

The relationship between human beings and computers can be divided into three main phases of development. The first "electronic brains"[4], as they were referred to in France, consisted of very expensive electronic components. Computers were therefore rarely used. They were controlled by a small number of actors and shared by several users. This was the era of the mainframe computer.

Progress in technology democratized the usage of computers to a point where they were designed for only one user. The golden age of information technology or the age of the PC had begun. The emergence of networks, mainly the Internet, whose advantages lie in their open architecture and set standards, have led to the creation of a wonderful communication platform used by individuals as well as objects.

In the 1990s, the idea of controlling an object from a distance with the help of the Internet, as well as allowing those objects to interact with one another, emerged. The concept of ubiquitous computing appeared. This new era is marked by the low cost of electronic components which paved the way for all objects to become "intelligent" machines since they dispose of their own computer.

Intelligent objects characterize this era in which information technology can be found everywhere and as a result becomes invisible to the user. The fact that information technology becomes invisible is based on two facts. First of all, it is miniaturized. However, subjective invisibility also plays an important role. Human beings tend to ignore devices that make their daily life easier. A good example of this subjective invisibility is watches. In the past, clocks were only available at important buildings such as churches or town councils. Now everybody owns their own watch.

This is similar to today's usage of processors which are integrated into many applications used in daily life (e.g. ERP, when programming a CD player or video

---

4 In 1954 IBM France was looking for a name to describe its electronic machine (IBM 650) which was used for the processing of information. They wanted to avoid a literal translation of the English term "computer", as in French this term was applied to machines used for scientific applications. The board of directors contacted Jacques Perret, an eminent professor and head of the faculty for Latin philology at the Sorbonne, Paris. On 16 April 1955, Perret responded to them, "What do you think of 'ordinateur?'". In his letter he explained that this word was correctly formed and used as an adjective to describe God who introduces order into the world. This type of word has the advantage that it can be easily used as a verb, "ordiner".

cameras). Just like clocks, they have become subjectively invisible. The user no longer "sees" them and is not very interested in them. It only matters to what extent they efficiently contribute to larger applications. Therefore, they are everywhere but the user is no longer aware of them. We are entering the third age of the computer, an era which some call the disappearing computer.

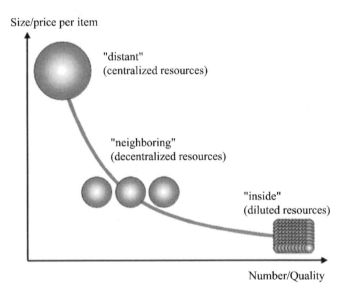

Figure 6.1. *A large number of small particles can be "diluted", i.e. sold at a low price. Permeated or diluted information systems are subject to this law. The real difficulty lies in the notion of a system that consists of small entities which form a collective intelligence or, in other words, a collective capacity to solve problems. These minuscule devices need to be able to communicate with one another to be able to solve a type of problem which otherwise could only be solved by a large centralized system. The challenge in the creation of these networks lies in their miniaturization and in the complexity of information that is passed between the individual and extremely small components*

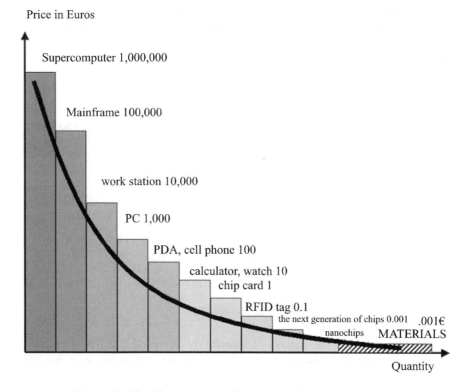

Figure 6.2. *The dilution process of systems and the impact on the price*

The development of computers follows a typical three-phase process of taming a technology. We have moved from an era where humans could only communicate with the machine in the computer's language and users therefore had to adapt their behavior to the computer, to an era where the user can communicate with the computer on an equal level. In this new era, machines will organize their activity around humans' needs and activities.

Moving from one era to the next happens gradually and applications from both eras tend to overlap each other. The first PCs on the market needed to be addressed in a specialized language similar to that of mainframe computers. Later, graphic interfaces, user-friendly applications, the mouse, etc. emerged.

This transition period is currently repeating itself. The user is surrounded by "intelligent" mobile devices (e.g. PDAs, cell phones, audio and video systems). However, all of these devices do not link with a central entity such as the PC. Today, it is still impossible to imagine an exchange between these devices. The model of intelligent objects that organize their activity amongst themselves calls the role of the user into question. Ambient networks are the next chapter in the history of information technology.

> The notion of a network has greatly evolved over the last decade. The idea of a network in which the users are in some form the terminals has been altered.
>
> The traditional idea of a network is based on conventional economic models which are based on a number of producers or, in other words, providers and a higher number of consumers. This asymmetric vision of a network is based on the idea that every consumer is a user i.e. a human being. With the miniaturization and democratization of technology, the number of potential nodes in the new generation of networks approaches the model of living organisms. Networks of objects allow for an information exchange between different machines without a formal hierarchy or given components with permanent roles. A dichotomy between the consumer and the provider no longer exists. Every node in the network can become co-provider or consumer according to the given requirement.

6.3. Highly diffused information systems – RFID

In the last 30 years the concept of traceability has made great progress in all economic sectors. However, this word still has not made its way into the dictionaries.

The idea of traceability (or product tracking) emerged with the use of modern industries and the need to identify components of products in domains where errors would have a strong impact on health and safety. This notion first emerged in aeronautics and later in the pharmaceutical industry. It is used as a means of controlling the location of a product during the entire time of its production process and later on its usage or the product's lifespan.

Every product or component of a product is tagged with a device that stores the information that affects the product during its lifespan. All important events are stored on this device. In the case of major events that might directly or indirectly lead to the destruction or major problems with the product, it is possible to retrace the origins of the problem. BSE in the 1990s, and more recently bird flu, have contributed to the fact that this application is now also used in the mass market.

The application of this concept allows for the storage and processing of high quantities of data. Digital information systems were therefore equipped with this kind of device very early on. Its usage, however, remained rather traditional. The devices stored the trace of the events in a rather traditional form. These were, for example, punched hole cards or barcodes which allow rapid identification with the help of an optical device which is used in combination with a computer.

The emergence of "intelligent" tags which could be traced via radiofrequency opened up new perspectives for the domain of traceability. This type of device, which is far from being a recent invention, has been miniaturized, improved and democratized over the last five years. Pilot projects with an economic focus were carried out in several different domains. The first standards were introduced mainly by organizations made up of the most powerful players in the industry and large service providers. Public reaction expressed fears over abuses in data protection.

> The principle of radio identification dates back to World War II. With the invention of radar systems the Royal Air Force was able to identify flying objects. However, a formal distinction between the enemy and ally airplanes had to be made. The Identify Friend or Foe System was introduced.
>
> This system of identification is based on a device called a transponder which was installed on board all airplanes. The transponder is linked to the radar which controls the movements of the airplane. The transponder creates a link between the radar and the airplane and receives an encoded signal and then automatically sends an encoded response back to the radar. This exchange gives a distinctive image of the plane on the radar screen.

Current RFID (*Radio Frequency IDentification*) systems use small tags made up of silicon or copper. These tags are also called Smart Tags or Radio Tags and the price per device is constantly falling. In 2004 a single tag cost between €0.10 and €0.15. Apart from its level of miniaturization and its price, this type of radio identification is fundamentally different from its predecessors as the power supply can be provided from a distance.

These tags integrate four different functions on a chip which is the size of several mm^2. These functions are a storage capacity of a significant size, a basic microprocessor, a communication device for small distances (which enables communication without direct contact) and a mechanism that produces energy with the help of an electromagnetic field.

The electromagnetic radiation of the read/write device on the chip creates an electric current that is induced in the coil-winding of the radio tag. The coil-winding is simultaneously used as a source of energy and as an antenna. This function lets the tag work without a battery, which allows for unlimited use at a low cost.

Currently, radio tags are used as anti-theft devices, for example, in supermarkets where RFID is used as a form of Electronic Article Surveillance (EAS). These anti-theft devices are, however, not based on RFID because they do not allow for the individual identification of a product. These are non-differentiated microbadges which consist of an antenna or transmitter that sends a simplified signal when subject to an electromagnetic field unless it has been deactivated.

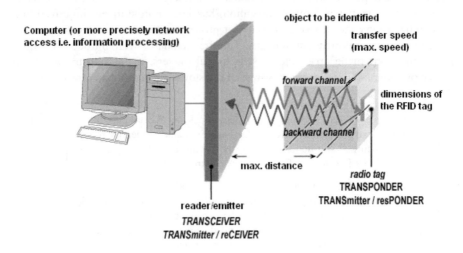

Figure 6.3. *Electromagnetic radiation from the forward channel creates an electric current which is induced into the coil-wiring of the transponder (RFID tag). This coil-wiring then functions as an antenna. The RFID tag sends a return signal via electromagnetic radiation (backward channel). The following 4 types of frequencies are used for the backward channel. These are from 125-135 kHz (60 bits, slow speed, reading only, transmission distance 10 cm), 13.56 MHz (2kbits and a higher speed, transmission distance 1 to 3 meters maximum), UHF 860-926 MHz (transmission distance 2 m to 0.5 W in Europe or 5 m to 4 W in the US) and last but not least 2.45 GHz which supports WiFi and Bluetooth (currently tested, 512 Kbytes, very high speed, transmission distance around 10 m). Lower frequencies, however, offer a better penetration into the actual material. In 2006 RFID microtags were estimated at being miniaturized down to the sub-millimeter level*

An RFID device materializes the properties of the product, i.e. it gives an image or a model of one product or several products. RFIDs can represent the products and their stored trace depending on the different levels of "intelligence". Radio tags are passive and enable simple identification only. Another form of these radio tags has a simple reading and writing facility which allows for new information to be written by the identifier. Smart tags are classed as intelligent and allow for information exchange between reading and storing devices. How sophisticated a device is depends highly on the type of microprocessor and the storage capacity of the chip.

The price of specialized microprocessors with an adapted performance is continually decreasing. Since 2003, the introduction of RFID systems has become a large-scale phenomenon. An increasing number of providers have joined the forerunners in this field. Texas Instruments, Philips, Sony, Alien and other producers of RFID radio tags had increased their production from some 10 million tags in 2000 to nearly half a billion tags by the end of 2004. The annual production lies at three billion tags in 2006 and is estimated at around 10 billion tags by the end of the decade.

The market could experience an upturn due to innovations. One of them might be the decrease in the price of chips which could go down to €0.03 to €0.05 per chip. The main users of this type of device are mainly the fields of mass consumption, especially the food sector. As the average price of a product in a supermarket is less than €5, radiotags need to be much cheaper in order to be profitable. Introducing this kind of device into the mass market does not make sense before it is cheap enough to be applicable to all products available.

The next generation of supermarket checkouts will enable consumers to pass a gate with their trolley which then recognizes the electromagnetic code of all articles and allows for automatic payment. The introduction of this kind of checkout is obviously impossible until all products are equipped with this kind of tag.

This was also the case with the system of barcodes[5]. This simple model is nowadays used as the basis for the supermarket or supply chain management which ensures the exchange between producers and distributing firms on a worldwide basis.

Compared to optical identification as is the case for barcodes, RFID, even in its most simple form, has two advantages. First of all, it allows for mass identification i.e. it can identify several products at the same time without being restrained by how it is passed in front of the reading device. This application is especially valuable during a marathon when many athletes reach the finishing line at the same time. Furthermore, the quantity of information passed on per code is much higher than that of optical identification. This information could be an identifying number which allows for the traceability of the product.

---

5 The barcode system, which will be replaced by RFID in the near future, is based on the optical recognition of a codification of either 12 of 13 numbers. In the USA, the Universal Product Code (UPC) is used while the rest of the world uses the European Article Number (EAN).

208   Nanocomputers and Swarm Intelligence

EAN 13 barcode for optical reading systems

(encoded in 13 digital figures)
The American UPC code is identical but does not take the first figure into account (value 0)

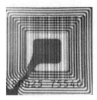

01 . 0000A89 . 00016F . 000169DC0

Header	ePC Manager	Object Class	Serial Number
0 to 7 bits	8 to 35 bits	36 to 59 bits	60 to 95 bits
country	company	product type	serial number

96 bits of ePC code on an RFID tag

Figure 6.4. *New identifiers called EPC (Electronic Product Code) use 96 bits for the decoding process. UPC and EAN use 12 to 13 different positions of decimal numbers (i.e. 12 or 13 numbers). The EPC code offers 9 to 10 different positions which allows for the identification of a product, a group of products or a product series. EAN and UPC do not dispose of this application which allows a product to be traced*

In the short term, barcodes will not be replaced by radio tags, i.e. EPC (Electronic Product Code). However, the majority of supermarkets plan to use RFID chips on logistic units which are, for example, used to move products (e.g. pallets) and on expensive articles (e.g. audio/video equipment, luxury cosmetics, alcohol, tobacco, etc.). However, the usage of RFID is still very limited and only applied to certain kinds of products. EAN and UPC will definitely be used parallel to the introduction of EPC.

> One of EPC's problems is its adaptation to the Electro Magnetic Compatibility (EMC)[6] standards which are valid in some countries (e.g. in Japan a frequency of 915 MHz is illegal). This standard, of course, has an impact on the performance of radio tags that are adapted to it. Their transmission distance can therefore vary according to the set standard. In Europe, these are 0.5 W and in the USA 4W for tags that use a frequency of 860-926 MHz. In the USA, this kind of tag is used in supply chain applications where the transmission distance needs to be of several meters (e.g. when unloading a truck outside).

---

6 Electro Magnetic Compatibility (EMC) describes a device that is able to deliver a satisfactory performance within its electromagnetic environment without producing an electromagnetic radiation that exceeds the given standards.

The price per unit of these tags is not the only problem challenging the general usage of RFID. Technical challenges are the main problem in the development of the "Internet of things".

6.3.1. *The Internet of things and the supply chain of the future – Auto-ID*

The idea is as simple as its application is difficult. If all cans, books, shoes or parts of cars are equipped with minuscule identifying devices, daily life on our planet will undergo a transformation. Things like running out of stock or wasted products will no longer exist as we will know exactly what is being consumed on the other side of the globe. Theft will be a thing of the past as we will know where a product is at all times. The same applies to packets lost in the post.

> The global association Auto-ID Center was founded in 1999 and is made up of 100 of the largest companies in the world such as Wal-Mart, Coca-Cola, Gillette, Johnson & Johnson, Pfizer, Procter & Gamble, Unilever, UPS, companies working in the sector of technology such as SAP, Aliens, Sun as well as five academic research centers. These are based at the following Universities; MIT in the USA, Cambridge University in the UK, the University of Adelaide in Australia, Keio University in Japan and the University of St. Gallen in Switzerland.
>
> The Auto-ID Center suggests a concept of a future supply chain that is based on the Internet of things, i.e. a global application of RFID. They try to harmonize technology, processes and organization. Research is focused on miniaturization (aiming for a size of 0.3 mm/chip), reduction in the price per single device (aiming at around $0.05 per unit), the development of innovative application such as payment without any physical contact (Sony/Philips), domotics (clothes equipped with radio tags and intelligent washing machines) and, last but not least, sporting events (timing at the Berlin marathon).
>
> The research on Auto-ID has taken an important step forward and led to the founding of two different organizations in 2003. EPC Global is a joint venture between EAN International and UCC is trying to industrialize the technology of Auto-ID. LABS, on the other hand, ensures the continuous research into the usage of RFID.
>
> The biggest challenge facing this program is the amount of information to be processed. It is reaching a level that was unthinkable in the era of traditional information technology.

If all objects of daily life, from yogurt to an airplane, are equipped with radio tags, they can be identified and managed by computers in the same way humans can. The next generation of Internet applications (IPv6 protocol; see section 6.4.1) would be able to identify more objects than IPv4 which is currently in use. This system would therefore be able to instantaneously identify any kind of object.

This network would not only provide current applications with more reliable data, as every object can be clearly identified, but would make way for new innovations and opportunities in the technological as well as in the economic and social sectors.

The Auto-ID program designs and promotes the components of this new network. In other words, Auto-ID establishes electronic codes for products (e.g. EPC that can be compared to UPC/EAN). Furthermore, it introduces a standard for tags as well as for reading devices which should be efficient and cheap (which means that a basic reading device should not exceed the price of around €10). Object Naming Service (OMS), a language used for the description of products based on XML (a product markup language), and an infrastructure of programs called *Savant* are also part of the association's efforts. Auto-ID develops certain standards for the identification process of objects which ensure that this kind of network will not depend on a standard introduced by a company producing RFID tags. Last but not least, the association is creating the infrastructure needed to allow the processing of the flow of information.

EPC is based on the idea that the only information carried by a radio tag is a reference. RFID allows for the creation of 'read/write' systems which would continually update the information stored on the chip.

However, the race towards miniaturization and reducing the costs of the individual tag favors a process which is "online" and used to update the trace that is produced. This trace consists of information that describes the article, such as the date of production, sell-by date, group of articles it belongs to, localization of the current stock, destination of the article, additional requirements, etc. All this information might take up more or less space depending on the format it is saved under and the nature of the product described.

The amount of data does not make the storing process any easier and currently the data cannot be stored on the tag only. EPC's aim, therefore, is to store this trace on servers that are linked to each other in a network.

In other words, only the EPC code will be stored directly on the tag. This code then enables the user to retrieve the required information on the object by using the network. The information will be stored on a server that is connected to a local network or the Internet.

A chip which is placed on an object as a form of identifier might indicate a URL on the Internet which then places the object into a hypertext and enlarges the concept of the hypertext to the Internet of things.

> The EPC is based on a reduced circuit that only consists of one Internet address per product that is saved on the RFID tag. This system therefore relies on an online database (Savant server or PML server).
>
> On the other hand, ISO circuits (ISO 18,000 norm), which represent traditional means of tracing a product (barcodes or paper files), have the advantage that the information is available offline and mainly stored on the tag instead of in a network.

ONS (*Object Name Service*) is a free localization program that allows for the identification of objects. ONS is the equivalent of the Domain Name System (DNS) which is needed to locate a specific content on the Internet.

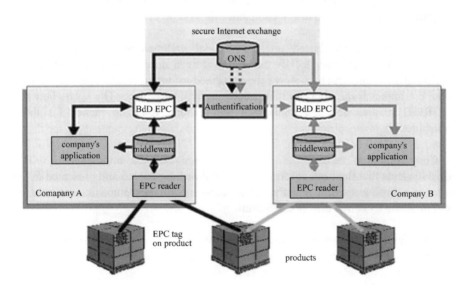

Figure 6.5. *Auto-ID, the concept of EPC and the supply chain of the future*

Organizations and companies that are exploiting the first generation of RFID systems will encounter one main challenge in the networks of identifiable objects. The amount of information that has to be processed will be extremely high and a burden to any company's IT system.

The Internet of things will encode 50 to 100,000 billion objects and follow the movement of those objects. Every human being is surrounded by 1,000 to 5,000 objects.

RFID readers deliver an enormous amount of information. Pilot projects in this field have allowed for an estimation of the volume of information that is generated if a company uses RFID to trace its products. The amount of information was 100 to 1,000 times higher than before. Another RFID based system traces all EPC events[7], i.e. all movement the product undergoes from point A to B, etc.

The introduction of RFID is subject to the same challenges as programs used in accounting, control programs used in industrial production, or even the administration of large telecommunication networks.

RFID applications are already capable of processing several terabytes of data, correcting errors, creating links between different events and identifying movement and behavior in real-time. The architecture of RFID is therefore inspired by these programs that have been in use in companies for several decades now.

These systems are based on a pyramidal architecture and are organized hierarchically. Every layer locally treats the received information (according to the level of abstraction) and then stores the processed information while waiting to pass it on to a higher level where further processing will take place. The layers between the RFID readers and the company's applications work as caches for those applications.

Communication takes place through message oriented middleware (MOM) which ensures that the basic information is validated, processed and presented in the form of a business event and therefore accessible to the companies information system. Auto-ID called this model Savant.

The Savant infrastructure is organized hierarchically and based on routers. Savant servers are installed in national database centers as well as in regional and local centers.

---

7 Auto-ID Center has introduced several short cuts. For example, only the last movement of a product is stored in the database.

Business Mutation and Digital Opportunities in the 21st Century   213

The RFID system can be compared to the nervous system of a human being. Its endpoints lie in shops, warehouses, factories or even delivery vans. Savant consists of three modules: EMS (Event Management System), RIED (Real-time In memory Data capture) and TMS (Task Management System). These modules ensure the acquisition of information, selecting from it and filtering it as well as classifying and storing it.

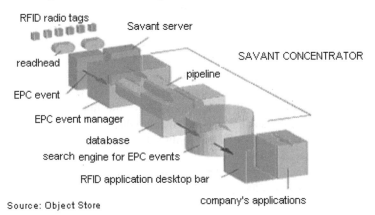

Figure 6.6. *Savant servers are components of a program that is being developed by Auto-ID and work as the central nervous system in an EPC network. The savant server receives information from the RFID readers, filters it and passes the information on to the company's information system or to the database. Savant servers provide the interface with the management units for events in EPC. A savant concentrator connected to a server allows for the intelligent regulation of the volume of received data*

Information that is relevant to products or objects is expressed in a specialized XML language called PML[8] (Physical Markup Language), which also represents an Internet standard. The Internet of things is an extension of the Internet towards the real world that could operate without any kind of interruption.

---

8 The usage of PML in combination with the appropriate vocabulary (PML Core and PML Extensions) allows for the representation of EPC objects so that they can be worked on by the entire system. They can be manipulated just as electronic documents that can be edited with the use of XML when co-operating with other companies. A program establishes XML schemata called DTD (*Document Type Definition*). They are compatible with PML and research is currently being carried out to find a way to integrate EPC and GTIN (*Global Trade Item Number*). Standards for the conversion between EPC code and optical representation EAN-13, SCC-14 and EAN-8 already exist.

6.3.2. *Economic opportunities vs. privacy protection*

"In 5-10 years, whole new ways of doing things will emerge and gradually become commonplace, expect big changes."

The shift from barcodes to the Internet of things and EPC makes the economic and technological domain focus on radio identification. Mass media alerted the population, who now regard new applications with increasing concern rather than being impressed by their performance.

Expect big changes! This press release from the Auto-ID and MIT in 2002 added fuel to the fire. RFID is clearly perceived as a risk and a major invasion of people's privacy.

The American tendency to strongly oppose any new risk that might invade their privacy is even expressed in their legislation. After the first application and pilot projects that involved consumers and statements which were too impassioned for technology and maybe even clumsy on a political level, a small group of consumers created a major association called CASPIAN (*Consumer Against Supermarket Privacy Invasion And Numbering*). The pressure of this lobby has led to several bills in this domain that mainly concern the deactivation of radio tags once the consumer has left the supermarket.

The fight between consumer associations and major companies is aggravated by a sensationalist press that looks for scandals in this field and whose information is not always very exact. As reactions and problems in this area have been extremely common in past years, it will not be possible to explain them all nor to give a summarized version in this book without moving away from the subject.

> Major commercial chains such as Wal-Mart are undeniably in favor of RFID chips. Even RFID's opponents agree with the usage of this technology for purposes such as stock taking and manipulations in warehouses, since RFID chips have clear advantages over traditional barcodes. This is why in the long run all suppliers will adopt the system of radio tags and EPC.
>
> However, all of these companies have always denied testing this technology on their clients. In 2003 a scandal that involved a large supermarket chain and Procter & Gamble was made public. The involved parties tested a marketing strategy without informing their clients.
>
> RFID chips were placed on lipstick. When a client chose a lipstick the chip set off an alarm which was then heard by company researchers that produced the lipstick. The researchers could follow the client through the supermarket with the help of cameras that were installed throughout the entire shop. The client's shopping, gestures and all other forms of behavior were taken into account and saved in a database. This survey was carried out during the whole time the person spent in the supermarket.
>
> Even though this observation was very indiscreet, it is important to note that the chip was not connected to a database nor could any personal information be retrieved by the researchers. However, the population began to fear that this kind of practice could be abused.

Without entering a passionate debate on ethics or giving a technological presentation, we should ask ourselves to what extent radio identification threatens personal freedom?

How important is the traceability of a bottle of lemonade and how is this linked to the protection of privacy? Apart from a group of sensitive objects, this type of information is rather insignificant for our privacy as this data is used to trace objects and not people. However, once it becomes possible to trace thousands of objects in a human's environment and create intelligent links between them, a profile of the person can be created. The creation of this profile is an intrusion into someone's private life.

Compared to other devices such as cell phones, GPS or video, the problem with RFID lies in its continuous communication which cannot be influenced by the user. Data is continually transmitted from a distance without any influence from the person carrying the object. This form of interaction cannot be stopped. An RFID chip is active for an unlimited length of time depending on its form of power supply. Consumer associations have mainly fought for the consumer to have the right to deactivate the system.

However, for certain applications a possible deactivation would mean that these applications no longer make sense. This applies, for example, to applications which focus on security, anti-theft and pirating. Parallel to the development of new applications, it is important to see to what extent they comply with legislation. The French law on freedom and information technology of 1978 can, for example, be combined with the RFID and privacy protection.

From an economic point of view, RFID is a fantastic opportunity even if it may evoke images of Big Brother.

Security might be one important domain for applications. Anti-theft systems as previously mentioned have already been used in self-service shops for a very long time. The possibility of identifying the location of every object makes the concept of theft impossible. Tests carried out in the field of car theft with systems based on permanent geo-localization gives a positive example. The miniaturization of tags and the strong decrease in their cost could allow them to be integrated into the next generation of bank notes that could then be traced. The principle of tracing objects and not people also applies in this case. The European Central Bank has started research in this field and will soon be testing pilot projects.

In a world dominated by the threat of terrorism, tests are being carried out on equipping access badges with radio tags. Localized and selective applications are becoming more and more diversified. Again, a person's privacy is not affected as long as this person is informed of the consequences that wearing this badge in a certain zone will have.

In sensitive areas such as airport security, applications could combine RFID badges and biometric applications in zones with limited access. The use of the two systems in combination with one another allows for the allocation of a badge to a person and checking their identity. The badge can only be taken over by another person if it is re-read. The reading process is connected to a database that gives the person's photograph and a detailed list of which areas the person is allowed to access. In 2003 the security system of the Chinese Communist Party's Congress was based on this type of technology. All major airports are currently considering this type of application. The European project Schengen Information System II[9] will also use this type of system in the next generation of passports. ICAO (*International Civil Aviation Organization*) has also carried out research on this kind of passport that could be read in three seconds from a distance of around 10 cm. Apart from a photo, which is traditionally used to identify a person, the new generation of ID will also be equipped with some biometrics such as digital fingerprints or retinal prints.

---

9 SIS II (Schengen Information System II) will allow for the storage of digital fingerprints and facial images (biometrics).

> An iris scan could be used as a form of identification which distinguishes all human beings on the planet. The image of this iris could be linked to a URL which would give a homepage with all information, basically an online passport. The iris could be used as a natural indicator for the database. This system is currently being researched for the next generation of passports, and is already being applied in zones with an extremely high security standard (e.g. centers for military research, data processing centers).

The health sector is another field for possible RFID applications. In 2004 the American FDA (*Food and Drugs Administration*) approved the implantation of radio tags under people's skin for medical purposes. Of course, the FDA was criticized for this decision as it is an invasion of privacy. On the other hand, the technical process may save lives, which is why the intrusion of people's private lives is more readily accepted. The FDA's decision will have a massive impact on companies like Applied Digital Solutions that produce this type of radio tag for the mass market.

The RFID implant is the size of a grain of rice and is implanted under the skin at the triceps between the shoulder and the neck. An intra-dermal medical syringe is used for this procedure which is almost entirely painless. The RFID chip connects the person to his/her medical file. The information can be consulted online on a secure connection. The identifying code on the radio tag consists of 16 numbers that can be read from a distance by, for example, passing a scanner over the area where the chip is implanted.

In the case of an accident, nurses equipped with such a scanner can directly establish a person's blood type (e.g. in the case of a blood transfusion) or be aware of allergies towards a certain type of medication.

However, health risks are not the only reason to accept the implementation of such a device. Members of the Baja Beach Club in Barcelona, which tried to offer a special identification process to its VIPs, were offered a free implantation of the chip *Verichip* by Applied Digital Solutions. The chip can be used as a form of payment when inside the club. The reasons for members accepting such an invasive procedure are based on an elitist system. An entire generation of teenagers may ask for this kind of indicator as a concept of social recognition. Is this a rather worrying step towards Orwell's dystopia?

> A technology itself is never intrinsically bad. It is its usage that may be so.
>
> "And now having spoken of it, I must add how charming the science is! and in how many ways it conduces to our desired end, if pursued in the spirit of a philosopher, and not of a shopkeeper!" (Plato: Republic: Book VII)

6.4. New challenges for web applications in a global network of objects

Ambient networks are without any doubt the corner stone of the next generation of information technology. This type of network ensures the communication of traditional mobile entities (e.g. smartphones, PDAs, different sorts of terminals, routers) and objects of a new generation such as intelligent RFID, biochips and interconnected microsystems. Whether silicon or nanotechnological alternatives are used, this concept is essential for the information technology of the next two decades. Nanotechnologies will broaden and refine these networks of objects to the maximum.

This type of architecture shows the problems that are encountered when using a mobile Internet. These problems are mobility, security, confidentiality, a stable service and the quality of service.

On top of these challenges concerning the network's infrastructure, new concepts are required since "intelligence" is no longer centralized but can be found throughout all minuscule objects. Anthropocentric concepts have always led us to the production of IT systems in which data processing, control units and calculating forces are centralized. These centralized units have continually increased their performance and can be compared to the human brain. The model of the brain has become the ultimate vision of computers. Ambient networks of intelligent objects and, sooner or later, a new generation of information systems which are even more diffused and based on nanotechnology, will profoundly change this concept. Small devices that can be compared to insects do not dispose of a high intelligence on their own. Indeed, their intelligence can be classed as fairly limited. It is, for example, impossible to integrate a high performance calculator with the power to solve any kind of mathematical problem into a biochip that is implanted into the human body or integrated in an intelligent tag which is designed to trace commercial articles. However, once those objects are interconnected they dispose of a form of intelligence that can be compared to a colony of ants or bees. In the case of certain problems, this type of intelligence can be superior to the reasoning of a centralized system similar to the brain.

This new vision of information systems does not only call communication platforms and the infrastructure of networks into question, but also imposes new algorithms that change the codes of traditional programs. This is why jobs in IT will also undergo dramatic changes.

6.4.1. *Complexity and efficiency of very large infrastructures*

How can complex and heterogenous networks be made to work efficiently on a large scale? Interconnected networks consist of billions of nodes; they dispose of an impressive variety of applications, infrastructures and communication protocols. All of these are challenges for telecommunications and new infrastructures. The main challenges can be sub-divided into three fundamental issues. These are (1) security, (2) quality of the service and (3) the size and number of the interconnected objects.

> In 1978 Minitel worked at the speed of a snail or, more exactly, at 1,275 bits per second. At the end of 2005 a DSL connection allowed for 25 Mbits per second. The speed was multiplied by 20,000 in 27 years.
>
> Metcalfe's Law states that the value of a telecommunications network is proportional to the square of the number of users of the system (n^2). Each time the number of users is doubled, the value of the network is increased by four.
>
> In around 15 years the Internet has moved from a pilot project to one of the most common technologies. In 2005 one in six people on this planet had access to it. The Internet has become an enormous digital network that has progressively integrated the domains of IT, telecommunications, TV and other cultural media.
>
> With mobility and very high speed, the Internet plays a central role in digital diffusion and mirrors the real world as every real object is associated with a virtual digital object. In a similar time period, the Internet that connects people will have evolved to an Internet that connects objects and mirrors the real world in the digital realm. Every object in use will be permanently connected to the Internet. The notion of a connection therefore remains transparent as objects spontaneously interact with one another without any human intervention.

When it comes to security, currently used mechanisms which are based on trust and encrypting processes that need a high calculating power are no longer adapted to ambient networks, especially to wireless ones. A wireless network might be used to obtain private data. The main problems in privacy protection lie in the traceability of mobile digital objects and the identification of the user while still respecting privacy, i.e. anonymity.

The technical bases for digital security were developed in the 1970s; however, these systems are no longer adapted to currently used systems as their scope is fairly limited. Controlled access to a system can only be employed with a small number of users who can be continually observed in order to detect abuse. This technique is therefore incompatible with networks of objects or complex ambient networks. In very large networks, not only does the number of objects and users increase greatly, but also the sensitivity of data and the required level of data protection. Access to bank accounts, for example, allows for the theft of money. The same applies to legal and medical information that can be extremely sensitive. Security in IT is based on one key technology: cryptography. Cryptography is an essential part of current information technology and allows for the protection of the user's anonymity just as it can prove the user's identity.

Currently used cryptography does not actually provide the level of security that it claims to offer. The majority of systems have not been installed by cryptographers but by engineers who often do a fairly poor job. Hackers do not follow the rules, they cheat. They attack the system in places and with techniques that were not anticipated. Their strategies can be compared to a burglar who cuts a hole into a thin wall rather than forcing his/her way through a security door. Hackers detect any kind of weakness in the system to gain access and steal, corrupt or act surreptitiously. Hackers are indeed more likely to win. Those who need to protect their systems need to take every weak point into consideration. These are constantly increasing, especially in the network of objects where they multiply exponentially while the hacker only needs to find one single weak point to enter the system.

This security problem will call into question the fundamental elements of today's information systems. To give an example, we can consider that the Internet consists of a number of computers that are interlinked with one another in a fairly flexible and open way. In order to connect another computer to the network, it is sufficient to use the standardized IP protocol and undergo some basic procedures to obtain an IP address. The concept of security, as previously mentioned, is based on trust and the implementation of cryptography which requires a large amount of calculating power. Introducing a security system into this type of Internet would require the installation of security devices that identify the user, control the exchange of data and observe the user if need be. This approach would strongly alter the open and flexible concept of the Internet. The security of the Internet of things has not been specified. However, a centralization of security devices at the heart of the network is a very likely solution that is currently under discussion. However, this security system could only be used for a limited number of systems and devices as it would be very easy to attack such a system in a large and very complex network. This solution can therefore not be applied. The security systems of ambient networks and the high number of interacting objects therefore need to be placed in the network gateways or interfaces. Every computer or object needs to be aware at all times of what kind of data is being exchanged and in which context this is done, as well as which level of trust is present between the two interlocutors. Security systems need to take into account that the architecture of networks is continually opening up and new challenges for security appear at the same time. Furthermore, the extreme capillarity of the network and the increasing mobility of objects play an important role.

The quality of service is another criterion that needs to be fulfilled when aiming for an efficiently working complex network of heterogenous objects. This notion, however, is not a new idea. Originally, the Internet was a best-effort network, which means that it did not give any guarantees concerning the delay of delivered information nor of the bandwidth. With telephone applications, video conference and other forms of communication, delay can no longer be tolerated. The IP protocol was extended with a system called Quality of Service (QoS) which gives priority to some data and earmarks a bandwidth for some types of data. This system is also used when there is an overload of data in the network. Originally, the IP protocol was introduced to use all forms of devices and still create a common network that was continually increasing. This concept was called *IP over everything*. Since all objects are interconnected, they create the Internet.

> This is why "intelligence", or in other words TCP applications of diffusion, are moved to the nodes in terminals rather than in the switching units. In this way it is easier to reprogram these applications. All devices have their common properties within the network without favoring their own characteristics. Independently, they are not very powerful, but every single one contributes to the interconnection. The resulting mechanism refers to best-effort systems rather than a QoS.

Once a packet of information, whether it contains image or voice, has entered the network it is mixed with other packets of data. In the traditionally used Internet, none of those packets has an indicator that gives it priority over other packets. This is why it is impossible to guarantee that they arrive at their destination in the right order and without any delay. In other words, if one word is shortened by a syllable or the first syllable of a word is stuck to the next syllable of the following word, then a telephone conversation would rapidly become unintelligible. This type of Internet is fundamentally different from a traditional telephone network. In a traditional telephone network, a telephone line is blocked for the duration of the conversation.

In order to solve this problem and guarantee a QoS, the producers of network equipment labeled the packets carrying voice or video information. The packets moving through the network are classified according to their content. This concept is known as Type of Service (ToS). However, all devices within the network need to be able to understand and use QoS/ToS mechanisms. This is not yet the case. Classic equipment, such as a DSL router, is often incapable of taking these labels into account which is why certain routers integrate the special function of IP telephony. However, other solutions exist. It is, for example, possible to earmark several dozen Ko/s of the bandwidth from a phone line for the entire duration of the telephone conversation. In reality, however, this strategy is rather unreliable. It can only be applied locally and does not have priority over other intermediaries in the communication process.

With the launch of Voice over IP (VoIP) in 2004 and television broadcast over the Internet in 2006, QoS has become the target of a large market with hundreds of millions of users in the near future.

However, for a future global Internet of things, QoS is far from being guaranteed. Since this network is so much larger and so much more complex, it calls into question this type of solution.

Since 1994, the IETF (*Internet Engineering Task Force*) has been working on networks that connect tens of billions of objects. The Internet was firstly a military application that made its way into universities, industries and governmental services in the USA in the 1990s. The rest of the story is very well known. At the beginning of the 21st century, the Internet has become one of the most used forms of media and consisted of more than one billion private users in 2005.

The convergence of the computer, audiovisual devices and telephony leads to the fact that all devices can be connected to the Internet. The number of devices accessing the net exceeds the number of IP addresses initially planned by the IP protocol in its currently used form, known as IPv4. In 1978, the inventors of the protocol believed that only several thousand computers and dozens of networks would need IP addresses.

> In IPv4 an address consists of 32 bits, i.e. four bytes. All possible combinations allow for 4,295 billion addresses that are theoretically available. This number of addresses would correspond to the requirements of the near future. However, the process in which IP addresses are distributed wastes these addresses and the possible stock therefore does not exist.
>
> Class A addresses, which are used in 125 networks and 16.7 million nodes are already all allotted. The same applies to class B addresses which make up 16,368 networks and 65,534 nodes. They are used by certain providers and industries. Class C addresses are earmarked for small organizations and small providers. Even all class C addresses are entirely used up.
>
> Different methods that allow for a more efficient distribution of addresses have been invented to counterbalance the weaknesses of the currently used IPv4 protocol. The most commonly known is called NAT (*Network Address Translation*). This approach combines a private network with a router or firewall that has only one public IP address. This address is used for a several computers. On the other hand, this approach is incompatible with the idea of a global network of objects that states that every object has its own address and is directly connected to the Internet. Furthermore, this approach is in conflict with certain applications of secure protocols.

The number of computers and networks that are interconnected is extremely high. Very large routers are needed and the work of network administrators is continuously increasing. Furthermore, we can observe a certain slow-down of the packets that enter the system. Even more powerful and faster routers, which are therefore much more expensive, are needed to ensure an efficient service. The main routers in the currently used infrastructure have around 70,000 routes each.

However, apart from the lack of IP addresses there are other important problems in the currently used IPv4[10]. IPv4 was created to transmit information but it does not offer any form of security. The fragmentation of packets increases the risk of overloading the network and slowing down the transmission of data. Furthermore, the mobility of every object in the network without changing its IP address is entirely incompatible with the initial protocol.

---

10 The increase in the number of users is not necessarily responsible for the slow-down in the transfer of data. IPv4 is divided into smaller networks which manage certain geographical zones. If, for example, 194.xxx means Europe, a parcel sent from the USA will be sent directly to Europe where its final destination will then be established. This system can be compared to the post. A letter is first sent to France, where it is transferred on to the region of greater Paris, then the city of Paris and finally to somebody's house.

The protocol known as IP new generation (IPv6 or IPng) came into use in the mid-1990s. On top of limiting the IP addresses, it offers a higher flexibility to resolve new problems that occur due to the introduction of large networks of mobile objects.

IPv6 allows for the interconnection of several dozen billion mobile objects that communicate with one another without the strains and inefficiency of the currently used IP addresses. IPv4 allows for 2^{32} or, in other words, 4.29 billion addresses, out of which a high number are wasted. IPv6 allows for 2^{128} addresses, which is a nearly unlimited stock of 3.4×10^{38} addresses. Furthermore, the header of packets that make up the packets of data is reduced to only seven fields for IPv6 instead of 14 for IPv4. The new protocol allows routers to process packets more rapidly. This is how any overload in the network can be significantly reduced.

However, the generalization of the communicating object[11] also introduces new options. The strain of obligatory fields in the header of IPv4 does not exist in IPv6. The context decides whether optional fields will be filled with content or not. The routers can therefore ignore options that do not concern them and process the packets more rapidly. As IPv6 disposes of functions such as identification and confidentiality it can be referred to as the predecessor of ambient networks of objects, which represent the next generation of information technology. The ever increasing chain of communication invented by humans can therefore be managed more efficiently and with a higher level of security.

6.4.2. *From centralized intelligence to swarm intelligence – reinventing the programming code*

As previously mentioned, the main challenge in large networks of communication objects which are based on swarm intelligence is their infrastructure. Their infrastructure is subject to strains of mobility, security, continuity and quality of service.

---

11 IPv6 is generally associated with the idea that one address is allotted to every object. What seems simple in theory turns out to be rather tricky in practice, especially when it comes to networks of communicating objects. If a stable address is given to a mobile object, depending on its location, interaction with it may become problematic. This is a routing problem. If, on the other hand, the address is dynamic and adapts itself to the given environment a binary decision diagram (BDD) needs to be used. In other words, the next generation of networks based on IPv6 requires changes in the protocol that are not available at this point.

The second challenge lies in the usage of ambient systems for new applications. The concept of programs with a high security standard that use highly distributed algorithms needs to be questioned, and co-operation tools used amongst micro services need to be miniaturized to the atomic level.

In ambient networks, processors and their corresponding programs become part of the objects and move away from the traditional image of a computer. These potentially minuscule objects need to be able to process high quantities of unstructured data. Their work needs to be carried out in an environment with hardly any stable connection which leads to a lower level of security.

This type of information system requires a new approach towards programming which takes into account new challenges such as the extreme variety of equipment, mobility, radically new data formats, less stable connections and the obligation to certify a program. This type of programming needs to be very intuitive since it needs to create interfaces that can be used by everybody, including young children.

This new type of programming enables the software engineer to write programs for more or less intelligent mobile devices. In other words, it allows the user to exploit data and services that are available on the Internet. In order to do so, formats that were developed to structure data need to be integrated into programming languages. The small size of objects and the architecture in which they are embedded led to the following requirement: the complexity of the algorithms used for the processing of information needs to be reduced. As the components are mobile, the programming code needs to be able to adapt itself to different types of environments. This flexibility requires an extremely high reactivity from the system which has to be able to recognize hostile components within it.

The creation of a program that can easily deal with the processors spread throughout the system and at the same time access large databases is an important step towards the next generation of information technology. However, the step from tests carried out in a laboratory and programs produced on a large scale adapted to the public requires a profound change in computer science. The transition of currently used systems towards omnipresent and complex applications which are used by the public and the industry goes hand in hand with a new type of programming that is based on elements from a database. As these elements stem from an intuitive language they can be modified, enriched, put together and even shared with new objects as they can access the same database of programs.

This kind of omnipresent digital universe is a perfect model of a decentralized architecture. This kind of system is so capillary that it is decentralized to the infinitesimal level. In a decentralized architecture programs mainly run parallel to each other. These programs do not work in isolation from one another and are able to access distant services or information. At this level, decentralized information systems need to be able to use resources with the help of calculating forces that are spread out in the network and can be located according to the calculating power engraved on the grid (see section 4.3.3). Programming languages need to be adapted to this. Programming languages such as Java that emerged at the beginning of the 1990s cannot be used very effectively for this type of parallelism. The threads or parallel running processes suggested by the inventors of this system are not integrated into programming languages. Furthermore, these mechanisms are not compatible with the applications of decentralized programs as they rely on pre-emptive scheduling[12]. This kind of scheduling might indeed reduce efficiency as the task that is carried out might lose the processor's capacities in favor of another task considered more urgent.

Life cycle of a thread

When a new thread is created, it is placed in the priority queue of the scheduler. During the next dispatch, the scheduler decides which jobs or processes are to be admitted to the ready queue. The ready queue contains all threads requiring the central processing unit (CPU). They are classed according to their priority. At every dispatch, the scheduler takes a thread with the highest priority and increases the priority of the remaining threads. This procedure ensures that all threads eventually move on to the CPU.

A programming language like Java still has, without a doubt, a bright future ahead of it. It is associated with a form of programming that uses distributed components and communication that is not synchronized. This application is called grid computing and refers to applications that are carried out in the context of the users' mobility and the calculating codes.

---

12 This kind of scheduling might indeed reduce efficiency as the task that is carried out might lose the processor's capacities in favor of another task that is considered more urgent. On the other hand, it is not referred to as pre-emptive since a task that is carried out cannot be interrupted by a task that is considered more urgent.

The notion of a component of a program marks the beginning of the industrial era in computer science. Just like industrial production, the field of programming has established elementary parts in the code that are interchangeable. This kind of principle has lead to the vast success of the automotive and electronics industry in the 20th century. The industry therefore firstly needs to develop databases of units of programming code that can be used for multiple applications. Object-orientated programming (OOP) creates a model that consists of a group of elements taken from the real world, a domain. Another group of attributes, in other words abstract entities that describe the object, is also used. Last, but not least, methods that describe the object's possible reactions towards its environment are integrated into the programming code. An object is a coherent entity that is made up of data and a programming code that works on this data. In 1967 Simula was the first programming language to introduce the concept of classes. A class represented all attributes and methods that describe an object. In 1976 Smalltalk implemented concepts such as information hiding, aggregation and heritage. Those are the three factors that describe an object's power.

In the 1990s, the principle of recycling certain objects was taken a step further. The modular approach and the capacity of recycling certain components of the source code are no longer sufficient. Componentization is the origin of distributed architecture that at this time forms the principle of codes. Programmers can now assemble macroscopic components of code to create a program without having to entirely write the code themselves.

However, the functioning of applications that create links between a large number of components, i.e. small entities of code in an open and vast space, has become extremely sensitive and complex. In this context, a program can no longer run at the synchronized timed work cycle of a traditional computer. In other words, it has become impossible to wait for one program to finish its task before carrying out another operation. When a program that is currently running needs to activate another program, their joint operation works in a non-synchronized mode. The program is not interrupted by waiting on the result of the program that it activated but continues to work on other tasks. The result obtained by the second program that was activated by the first will be integrated into the first once the second transmits it.

Programs need to be adapted to a vast network of tiny and heterogenous objects that represent the next generation of distributed capillary systems. These networks will be based on the architecture of non-synchronized communication and the notion that parts of programs can carry out tasks independently. However, the next generation of these components also offers the capacity of composing themselves in a natural way, i.e. according to the behavioral semantics and their operational efficiency. An almost "genetic" code ensures that the components are aware of who has the right to do what on a certain object. In this way, global security is reinforced and risks of information leaks are diminished. This is a very good example for an auto-adaptive approach.

Nature has given us several examples of how minuscule organisms, if they all follow the same basic rule, can create a form of collective intelligence on the macroscopic level. Colonies of insects perfectly illustrate this model which greatly differs from human societies. This model is based on the co-operation of independent units with simple and unpredictable behavior. They move through their surrounding area to carry out certain tasks and only possess a very limited amount of information to do so. A colony of ants, for example, represents numerous qualities that can also be applied to a network of ambient objects. Colonies of ants have a very high capacity to adapt themselves to changes in the environment as well as an enormous strength in dealing with situations where one individual fails to carry out a given task. This kind of flexibility would also be very useful for mobile networks of objects which are perpetually developing. Packets of data that move from a computer to a digital object behave in the same way as ants would do. They move through the network and pass from one node to the next with the objective of arriving at their final destination as quickly as possible. However, just like ants in the natural environment, they might encounter obstacles. If an artificial ant is held up at a node because of too much circulation, it places a small amount of pheromones on this node, while those ants which are moving quickly as they are not affected by too much circulation mark the way they are taking with a lot of pheromones. Due to this information other ants use the faster route. This is how the network regulates itself without any need for intervention on a macroscopic level.

Colonies of insects develop swarm intelligence that has seduced software engineers. For a very long time, entomologists have neglected the study of ants as they were perceived as being at a low evolutionary stage in the insect world. Today, colonies of insects and their collective behavior form a vast field of studies aimed at problem solving in complex areas such as robotics or information technology. As an example we will now take a look at virtual ants and artificial pheromones used to combat information overload and congestion in networks.

The main quality of the colonies of insects, ants or bees lies in the fact that they are part of a self-organized group in which the keyword is simplicity. Every day, ants solve complex problems due to a sum of simple interactions which are carried out by individuals. The ant is, for example, able to use the quickest way from the anthill to its food simply by following the way marked with pheromones. The individual's orientation is easy, but its capacity to choose between two options remains rather limited.

When a colony of ants (i.e. several hundred thousand individual ants) is confronted with the choice of reaching their food via two different routes of which one is much shorter than the other, their choice is entirely random. However, those who use the shorter route move faster and therefore go back and forth more often between the anthill and the food. Progressively, the level of pheromones increases on the shorter way. This is how the message is transmitted and the ants end up using the shorter way. Circulation is therefore regulated to an optimal level.

The notion of swarm intelligence is based on small agents that individually dispose of only very limited resources (e.g. intelligence, mechanical force, etc.). Co-operation remains very basic and, paradoxically, as a result highly efficient. The tasks given to the individual are very simple. These are, for example, pick up an egg and move it to a safe environment, take food from another individual and move it to a specific environment. This behavior is rather primitive but, as a whole, the results are very rich and very coherent. The difference with human social models lies in the behavior of the individual which, in the case of colonies of insects, only serves the interest of the community, while humans are social animals out of pure necessity. Free from any kind of personal interest an ant would not hesitate a moment in sacrificing itself to increase the advantage of its swarm as a whole. Their choice is based on a simple gain/loss principle. This type of decision-making is very fast and based on simple criteria that are established for the entire community.

A large number of very small individuals have one other important advantage. When it comes to the relation between power and mass in the fields of intelligence or mechanical force, the results are greatly increased. The difficulty lies exclusively in the efficient coordination of the dispersed know-how.

The basics of programming code of the future, i.e. diffuse application codes, are based on three main principles. First, the interaction between the codes of two objects becomes weaker as the number of objects increases. Non-synchronized communication is therefore the future of programs based on swarm intelligence that run parallel to one another. Secondly, the notion of microcomponents is strongly connected to the spreading of the code that is controlled on a macroscopic level. Last, but not least, algorithms need to adapt to certain problems, i.e. they need to find methods to solve problems themselves. Future programs will develop according to the task they carry out within their environment. The concept uses mutant applications.

Viruses that enter our networks are considered very dangerous if they manage to modify certain lines in their program by themselves so as not to be recognized by anti-virus software. Currently, no virus is able to modify the way it attacks a system in order to adapt itself to the system's defense mechanisms. This kind of mutation can so far only be observed with living organisms that mutate through evolution to adapt their genetic code to their environment. On the other hand, this type of adaptation is, at least, theoretically possible for computer programs.

Currently, algorithms that make up programs are limited to tests and instructions (which work in a loop) that allow for a problem to be solved systematically. An algorithm therefore needs to take all possible scenarios into account. This type of program does not evolve in any way and therefore only takes into account situations that someone programmed it to deal with. Chess players will improve their strategy by learning from errors and success. The best computers in this field do not perform much better than a human player as they need to play tens of millions of times in order to improve their performance. An impressive example is computer games for young children. The computer often wins the first twenty or thirty times until the child finds its weaknesses and wins nearly every other time. The capacity of distributed programs to adapt themselves to different situations is certainly a major challenge for future programming codes. However, evolution, whether it is natural or artificial, is based on the ability to adapt.

> Researchers at the University of Texas have created the predecessor of a mutant algorithm known as the genetic algorithm. This algorithm evolves each time it is used and takes on board the efficiency of each method used to find a solution to a given problem. This type of algorithm simply applies the rules of evolutionary theory. The best solutions are kept and multiplied while those approaches that prove themselves inefficient are no longer considered.
>
> The FBI tested this type of algorithm. They used it to compress around 50 million digital fingerprints in their database. The FBI previously used an algorithm known as *Wavelet Scalar Quantization (WSQ)*. This algorithm allowed them to reduce the space taken up by digital fingerprints and avoid the usual errors in the identification process led to by the compression of data. Furthermore, it was efficient enough to allow for 50,000 enquiries a day to the database with an acceptable delay i.e. the time needed to obtain a response.
>
> The idea to outperform WSQ was based on the fact that the "genetic algorithm" was only given basic instructions needed to compress the images. After compressing around 50 generations of images, the computer itself had elaborated a new method of compressing images. This method outperformed the WSQ algorithm which was created entirely by humans.

6.5. The IT jobs mutation

Performance and capillarity will be the two main themes of new information systems or, in other words, the digital world over the next decade. IT is no longer a discipline on its own, nor a branch of a company that caters for specific needs. Calculating power, miniaturization and the decreasing cost of the individual device have opened up radically different perspectives that were unthinkable in the 20th century.

IT and digital applications are, of course, a part of technology that describes the individual device. Until now, this book has mainly dealt with this type of technology. However, technology is only the starting point for new developments. The introduction of new developments relies, after all, on programming codes which will make devices work.

As technology is transforming there will be changes in the professions and the role they play in integrating new technologies. In other words, as devices change, the ways in which they are used also change. This is a principle that can be observed throughout the history of science as well as in technical development. The programming industry has changed with the emergence of distributed applications. Similar changes took place in companies' IT departments with the emergence of the

Internet. The changing architecture of programs has altered the way in which complex applications are designed.

The extremely high performance of new computers in ambient networks paired with the limits of how much code programming teams can write has led to the emergence of a modern approach to programs. DARPA has launched an ambitious project called HPCS (*High Productivity Computing Systems*). Between 2007 and 2010 they should provide the digital industry with new tools and components for ambient systems with a capacity superior to one petaflop. This project follows Moore's Law in terms of programming. The productivity of programming teams should double every 18 months.

The efficient development of complex applications will no longer concern only the size and cost. Mobility, capillarity (the capacity of minuscule systems to work autonomously as well as co-operate with one another using ambient intelligence; see section 6.4.2) and security, lead to the need for significant reforms in code development, maintenance and also in the way IT is used.

> Programs are becoming increasingly complex. Around 15 years ago important applications required billions of lines of programming code. Today, programs are often only several million lines long. The reduction in the number of lines often leads to a higher number of errors in the program. The National Institute of Standards and Technology (NIST), a US body in charge of IT standards, has estimated the cost of errors in programs at up to $60 billion per year. $22 billion could be saved if the testing equipment in the programming industry were to be improved.
>
> 700,000 engineers are employed within this economic sector whose turnover is more that $180 billion. 80% of the sector's resources are spent on the identification and eradication of errors in programming code. The sector's resources are, however, still insufficient to avoid the emergence of viruses.

The methods used to develop and maintain programming code have greatly improved over the last decade. The Internet and the principles of ambient architectures are the driving force for this development. The use of small units of code that can be recycled and are interchangeable has introduced an industrial dimension into programming and made it more efficient. The architecture of ambient networks has allowed for the delocalization of activities into regions where the cost of labor, etc. is more attractive. However, the transformation of professions still remains rather superficial.

As IT is subject to market forces, applications need to evolve permanently[13], their speed constantly needs to increase no matter how complex or interdependent they are. Due to ever increasing global competition, the aim of reducing the price of IT and the possibility for a product to be adapted or developed further at a relatively low cost has led to companies adopting a new approach to IT.

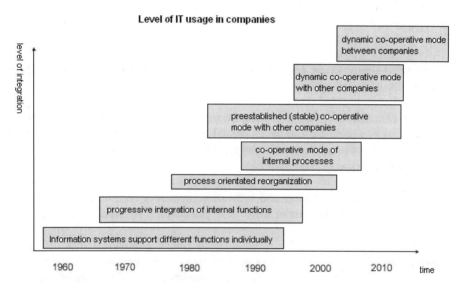

Figure 6.7. *With the emergence of transactional systems in the 1980s, the process-based approach replaced the traditional function-based approach. A company's internal processes interacted first of all amongst themselves within previously established schemata. Collaboration has made the exchanges more flexible; however, they remained within the previously established schema. At the end of the 1990s, dynamic collaboration paved the way for more flexible exchanges that no longer needed previously defined concepts within the system. With the beginnings of ambient intelligence this open form of collaboration is leaving the field of companies and is being applied to a generalized network of objects*

Furthermore, the notion of functions and processes that are restricted to a specific company or to an interface of a process that connects two companies is applied to a larger set of applications. Other players, or even external objects, are entering the playing field. They are end users, clients, suppliers or other players that have one of the following terminals: telephone, TV, laptop or even a moving vehicle, in which intelligent objects are integrated.

---

13 A good example is Google's beta mode: a stable application, but open enough in terms of functionalities, allowing continuous evolution according to the required functions.

These processes that progressively take their parameters from the real world are of an extremely short lifespan as they change people's daily life.

The development and maintenance of applications can no longer be reduced to a group of programs with a constant lifespan but needs to take into account the high number of parameters of systems that interact with one another. This immense diversity is replacing the boundaries that IT projects of the last century encountered. These boundaries no longer exist.

We are entering a new era of IT that will industrialize the development and the maintenance of digital applications.

In the next decade, a company's software engineer will no longer develop but recycle. This development can be compared to a mechanic who does not produce his/her own screws or an electronics engineer that assembles integrated circuits in an even more complex way.

The increase in web services and software components has already displayed this new trend. The new generation of software engineers does not start from scratch.

They consult online databases in order to find a component that offers the required function or a similar unit that might already be available. Instead of creating a new structure, software engineers assemble recyclable units of code that can be connected to one another and operate as a whole.

Software development is therefore being replaced by integration on the smallest level, which therefore could be seen as just another form of development. The design and maintenance of those databases that need to be sufficiently stocked is based on industrial programming code. These are standards and the extreme subdivision of functions. When mentioning standards and protocols, strains and increased complexity of the post-pioneer period must also be mentioned.

At first glance, the advantages of the recycling method may include making programming easier, but they are, without any doubt, only applicable in the medium term. Writing a code by using already existing components might at times be harder than inventing a basic code.

However, programming that is based on the use of already existing units is the only approach which is able to react to current constraints, i.e. omnipresent applications, post-modern applications that overlap each other, interact with each other and are of a short lifespan. At the same time, the following aims have to be reached: reactivity, productivity and localization at a low cost.

6.5.1. *New concepts in agile software development*

Programming codes have always been based on algorithms and the creation of models. To write a program, apart from the tests and instructions that make the computer run in the first place, a complete analysis of all possible consequences needs to be carried out.

In isolated systems which were used in the past, this type of analysis was easily applied. The lifespan of applications was very long and interactions with the outside world were not common. Often several months or even years were required to write the appropriate code and analyze problems, which, due to the previously mentioned facts, was not a problem.

The notion of interaction with the environment started with distributed applications and models used for analysis were progressively confronted with external strains that introduced new problems which did not comply with general theory.

Competition on the market has increased this pressure due to the rising demands in terms of access speed as well as for other specifications such as lifespan of the individual product, logistics and administrative delays. A solution to a problem, no matter how efficient it may be, does not make any sense if it is not made available in a day or, in the worst case, within a few weeks.

The notion of ambient intelligence and hyper-distributed models even increases the time pressure that now links a problem and its possible solution. Agile software development is a new approach in the field of programming. This approach favors a short delay instead of a complete analysis of the given problem.

This agile software development[14] allows for the client of a certain application to participate as much as possible in the way a program is designed. Previously, the client was only associated with the feature specification. Now there is a direct reaction to the user's needs. Agile software development is therefore much more pragmatic than previous methods. It is aimed directly at the users' needs and satisfaction instead of respecting a formal contract that was previously established.

Agile software development emerged in the 1990s and is marked by its capacity to adapt to a certain context and evolve constantly according to the changes in functional specifications. Agile software development is justified due to the high

---

14 In 2001 this method was formally agreed on in the Manifesto for Agile Software Development which was signed by 17 important figures in the world of business and information systems.

instability in the field of technology and the fact that clients are often incapable of expressing their needs with enough detail to make it possible to create an entire project. The first applications of agile software development still remain questionable on a global economic level. However, they have introduced a new form of co-operation between the client and the provider of a possible solution. Applications need to be built around the needs of individuals and their modes of communication instead of being based on certain processes and tools. It is no longer essential to deliver a final product but a programming code that can be adapted to the user's needs. Specifications in the contract no longer stop the continuous delivery of programming code which directly reacts to the changing functions. The code can be accessed directly while it is still being written.

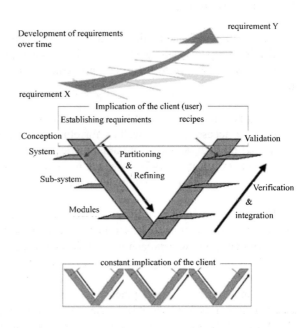

Traditional method

The client is only implicated in the initial phase, when his/her requirements are established and in the final application process. The client's needs may change during the productiob or development of the application.

Agile software development

Development of a basic version. More functions are integrated at later stage during an interactive process based on continuous dialog with the client and tests which are carried out throughout the life cycle of the product. The organization of projects based on this type of development is difficult.

Figure 6.8. *The clients currently not only own the application but they can also direct it. It is in this way that the production process works much faster and results can be obtained in several months as opposed to the previously required two to four years. The initial version of agile software development is constantly gaining new functionalities and is permanently adapting itself to new demands*

At the beginning the 21st century an even more radical version of agile software development emerged. *eXtreme Programming* (XP) is adapted to a small number of programmers (around 10) and flexible with regards to changing requirements.

Business Mutation and Digital Opportunities in the 21st Century 237

This type of programming has often been criticized for being excessive, i.e. the methods used lead to a rather heavy programming code that contradicts the basic philosophy of agile software development. On the other hand, XP uses the programming code of the future, i.e. hyper-distribution and adaptability.

The form of XP is also known as pair programming[15], a not entirely new principle. It consists of a synthesis of simplified usage and principles that are well known in the industry and used in project management. The time required to write a program is rather short, it only takes a few weeks. During an iteration phase, possible client scenarios are tested and determined.

The team uses these scenarios to establish which tasks have to be carried out and tests them. Every programmer chooses tasks and writes the code with his/her partner, which is why this type of programming is called pair programming. Once all functional tests[16] give satisfying results, the product is delivered.

The work cycle is repeated once the client transmits new scenarios. The first delivery is usually the program's longest part of code and therefore needs to be carried out over the course of several weeks.

However, after the first delivery, the iterations become shorter and are written in work cycles of around one week each. XP advocates control and a careful limitation of the client's demands.

Additional functions cannot be added at the beginning of the programming process and the application cannot be optimized before the end of the work cycle. The aim of iteration is to implement scenarios that were chosen by the client only[17].

---

15 This type of programming is carried out by two programmers. The first one called the driver uses the keyboard. This person will write the code. The second, known as the partner, helps by suggesting new possibilities and tries to find solutions for possible problems. The developers frequently change partners, which enables them to improve the collective knowledge of applications and improve the communication throughout the entire team.
16 Before adding a new functionality, the programmer writes a test program that verifies whether the functionalities work as intended. This test will be kept until the project is completed. At every modification all test programs are launched. All tests need to be carried out in less than a day. Possible errors therefore appear immediately.
17 Agile software development takes neither the form of a sample, nor processes, nor needs, in the conventional sense of the term. This is what agile software development stands for. This new approach of systems that continually evolve is a "nightmare for software engineers" to use the term that even IT directors use nowadays.

The idea that progressively changing applications is a waste of time is wrong, since what applies today (and even current outlooks for the future) might not be applicable in the future. Furthermore, an application which is not very extensive in terms of functionalities can be modified much more easily.

XP is far from going back to the origins of programming, i.e. writing an entire programming code in the way it was done before the industrialization of this sector. By shifting the focus towards correct programming practices, XP is a succession of short iterations which involve the client at each step of the programming process.

The relationship between the provider and the client is being defined in new ways. The results are an impressive quality of programming code, short time spans until delivery and high customer satisfaction[18].

Agile software development is, without a doubt, the programming code that creates ambient intelligence or minuscule objects that continually interact with one another and contain only the minimum of programming code needed to reply to specific requirements.

These specific requirements are less complex than current applications as they are localized. On the other hand, they are less stable as the environment continually evolves. Ambient systems are strongly connected to the outside world and are different to traditional applications that were isolated on purpose because of the fear of external threats that might affect them. This leads to problems as the outside world is constantly developing.

The development of every possible system is always subject to the life cycle of the bell curve. To solve a problem in the real world an abstract model is created.

This abstract model which represents reality will give a possible solution for the given problem. This abstract models is used in tests that are carried out to see whether the solution is appropriate.

---

18 In traditional companies or organizations, users are often hesitant about technological changes. Traditional users are often reluctant to employ technical innovations and will not ask for these developments to be integrated into their applications.

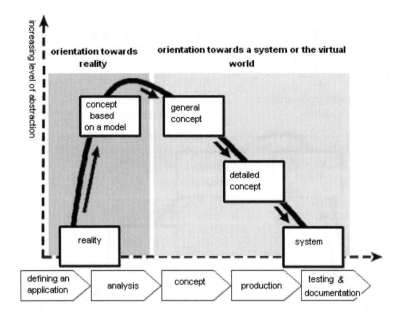

Figure 6.9. *The bell curve in system development*

IT applications have integrated a monotonously increasing level of complexity. In other words, systems are developed to respond to ever more complex problems. The level of abstraction needed to design those applications is therefore ever increasing, as is the time needed to find solutions for problems within a system.

In reality, the work cycles have not increased proportionally to the increase in the complexity of applications, but new tools and methods have improved productivity (e.g. automatic generation of programming code, test robots, etc.). Without a doubt, the level of abstraction weighs heavily and makes changes in traditional applications of programs rather difficult.

Diffuse information systems, under the umbrella of ambient intelligence, set a limit to the level of abstraction. Since the power of a processor is split up into minuscule systems that interact with one another instead of being concentrated in one large system, the processing capacity of the single system is curbed. Even an exponential increase in the performance of a single chip does not contradict this development.

Given their processing and storage capacities, applications embedded in capillary systems need to be of a simpler logic and anticipate situations that were not predicted and therefore introduced in their programming code.

Traditional applications take all possible scenarios into account: the most complex problem is therefore considered a priority. This is a rather complex system with a long life span.

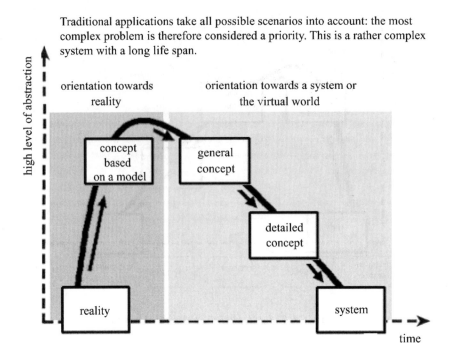

Ambient intelligence curbs the processing power of a single device. The level of abstraction i.e. complexity, is lower in an individual device. The level of "intelligence" is therefore lower in a single device, but a large number of these devices where each individual one reacts to local problems outperforms mainframe computers.

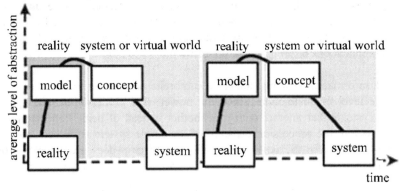

Figure 6.10. *Systems of ambient intelligence: a lower level of abstraction is reflected in shorter life cycles. The development is therefore very agile*

A lower level of abstraction leads to shorter life cycles, which is part of agile software development. Even if this kind of system is very basic compared to large programs that use intelligence to explore all possible scenarios and therefore take a long time to be developed, it is not less efficient. The simplicity of programs that are written rather quickly has the advantage of being up to date and being able to reply to the needs of a given system.

A high number of minuscule systems that are interconnected and address local problems and requirements might outperform a large system whose functionalities are more refined. "We invent operations which are less likely to be needed, but carry them out more often."

Ambient systems take fewer scenarios into account but they are more up to date and there is a higher number of ambient systems. They can therefore better adapt to their environment and their solution to a given problem is more substantial. A good solution to a problem that no longer exists in the given form is worthless, while a solution that is rather superficial but given more quickly is a very good option.

We are now left with the problem of coherence. How can it be guaranteed that this type of interconnected system produces an organized solution and does not transform itself into a sum of anarchic operations? Once more, biology and the study of colonies of insects show that it is unnecessary to impose a central control unit to obtain macroscopic results. Control systems in a swarm work parallel to one another and are coordinated by a principle which is just as universal as evolution itself. Those operations which turn out to be useful in a certain environment are kept and used again in the future. Operations that prove to have a negative outcome are identified and no longer carried out. However, this type of self-adaptation of systems, in other words their autonomous control, currently remains one of the most critical issues when writing programming code for ambient systems.

6.5.2. *Ambient intelligence and the delocalization of jobs in the IT sector*

During the last decade distributed architecture and the Internet have largely favored the movement of the delocalization of jobs in the IT sector. The productivity of information systems is a priority for most companies. What are the consequences if a new program which is distributed even further finally reaches the boundaries of possible interaction? Ambient intelligence might therefore be a threat to traditional jobs in the IT sector. What does this threat actually consist of? Is off-shoring[19] of

---

19 In the 1970s US companies were the first to introduce the off-shoring concept and move parts of their activities to countries with low labor costs. Continental European countries were

development and production in this sector a temporary trend that will be reversed once managing directors are made aware that performance does not meet the required standards?

This theory that clearly reminds us of the IT sector in the 1980s is not very likely. The first experiences with off-shoring have definitely suffered from the workforce's differing level of qualification emphasized by the language barrier. Only English-speaking countries could escape the Babel syndrome by delocalizing to India. The disastrous impression was reinforced by a large number of articles in the press on the situation.

The reality, however, was much less disastrous that it seemed. Of course, the time difference, which was often around eight to 10 hours, profoundly changed the way work was carried out. Problems have occurred in fields in which the physical presence of another person often gave the false impression that needs were taken into account and that solutions were offered more quickly. In reality, eight hours of time difference is nothing when it comes to specific requirements and delays for the delivery of applications that have been tested and are reliable. An increasing language barrier which replaces the ideal communication between, for example, two native English speakers remains problematic (e.g. co-operation within Europe). The theory that everybody speaks English is far from being applicable to the generation of today's managing directors and employees. However, this theory will be applicable in years to come.

When delocalizing, the quality of the workforce in the IT sector is not at all problematic. The contrary is actually true; centers for IT development certified at a CMM5 level are currently working all over India with a number of university graduates that can be compared to the number of university graduates working at Silicon Valley. The average level of education for software developers in Europe is a Bachelor's Degree, while Indian employees often have a postgraduate Master's or even a PhD. Some of them have even studied in the most prestigious American universities. However, they do not consider software development as a task which is inferior to, for example, project management.

Our vision of delocalization therefore needs alteration. Reasons for delocalization are changing. Previously, services regarded as too expensive or non-strategic, even if it meant that concessions had to be made in terms of the quality, were delocalized. This trend is changing and companies are now looking for

---

more reluctant to do so and only took this step in the last decade. In the field of IT it is principally the fields of software development and maintenance which are delocalized.

excellent quality. India, for example, has become one of the main centers of knowledge and competence when it comes to Java[20] programming.

> The *Capacity Maturity Model* (CMM) defines best-practice when it comes to software development and integration. CMM ranges from 1 (state of anarchy) to 5 (less then one error per 100,000 lines of programming code).
>
> CMM5 is therefore the highest possible standard and measures increases in performance in the categories of cost, schedule, productivity, quality and customer satisfaction. Only around 10 companies in the world reach the CMM5 standard. Most of these are located in India.

Offshore development has succeeded in the area that even famous analysts describe as a severe problem. Delocalization was first only undertaken to cut down on costs, especially in areas that require a large workforce. Delocalization means that the productions is decentralized and spread throughout the world. Production is moved to places were the workforce is cheapest and a lower productivity, which is compensated for by the lower price per single item, has to be accepted just like lower quality in some aspects of the product.

The delocalization of production is nearly the last phase in a product's lifespan. Only types of productions which have become so banal that know-how is no longer a competitive advantage for traditional Western companies are delocalized. The last aspect of competition is the price. Products which were previously considered as strategic or high-tech have reached the level where their technology or production has become banal, as was the case in the textile industry or later on in the production of TVs. In the 1960s and 1970s, the production of TV sets was considered high-tech and became progressively commonplace in the 1980s. Since the 1990s, with the phenomenon of micro computing, the production of computers has followed the same trend.

---

[20] The analysis of the situation is even more worrying for European countries when comparing the number of university graduates with a Bachelor's, Master's or PhD per year with India and China. In the 1960s, Korea had the same GDP as Afghanistan and was 21st in the OECD's list on the percentage of adults who obtained a degree. Today, Korea is third on the OECD's list when it comes to 25 to 34 year olds who are university graduates. Some European countries such as Ireland, Portugal and Spain have also improved their results. The main European economic powers such as France, Italy and the UK have only maintained their results.

The pervasion of delocalization lies in the continuous changes in production sites. The transfer of a production plant to a specific region that offers low labor costs adds value to this region and the workforce will progressively become richer. The population will therefore progressively reach[21] a higher standard of living and enter the inflationary vicious circle of labor cost. The company will then move their production to another region whose workforce is cheaper and constantly move their production.

Software development could have entered the same vicious circle. Luckily, this was not the case as the delocalization of software development that started with the same intention of cutting price quickly shifted its focus towards excellent quality. Countries like India and later on Mexico, Brazil as well as eastern European countries understood that lower price does not offer the perspective of long term growth but that quality and innovation are needed. While innovation still remains the domain of traditionally strong economies such as Japan, the USA and western European countries, excellent production and realization of specialized tasks has very much become the field of emerging economies. Today, software development and maintenance of applications is firstly delocalized to India because companies are convinced of the quality and the reactivity of the service offered. Cheap labor costs are no longer the only reason for delocalization.

To what extent might web diffusion amplify the trend of delocalization? Ambient intelligence is the successful outcome of distributed informatics and the interconnectedness of components. To design and organize a network of objects that can react locally and interact autonomously with other objects requires programs to be written for these objects. On a global level, the programming code for this type of network will be tens of billions of lines long. This quantity of programs, which are able to interact with one another and the relevant documentation, is unheard of today. An extreme discipline which is far from the creative processes of the early stages in programming is therefore required. Industry standards that lay out norms and quality standards are the only possibility for the creation of this type of organization. Due to the competence that has been developed in the fields of software development in some counties where production has been delocalized, there has been certain stabilization in the creation of applications. In the future, these developing centers will possibly become global players as they will produce the programming code of the future. India, Mexico, Brazil and eastern European countries have benefited from the delocalization process. The purchasing power and the standard of living in the regions have increased, but the production has not been

---

21 In a study carried out between 1939 and 1943, the psychologist Abraham Maslow defines a hybridization of human needs which is linked with different levels of motivation. Reasoning is a superior need that can only be carried out if all other basic needs are satisfied.

delocalized as these countries managed to combine competence and high quality standards.

6.5.3. *New opportunities for the profession*

In only half a century IT has moved out of the laboratory to become a general source of our daily lives and activities. From the 1950s to the 1970s, the days of IT's pioneers, this field was only a relatively marginal economic sector. During the course of these three decades IT controlled mechanical operations such as sorting mail and administrative documents or was used for calculations in laboratories that became the new temples of applied mathematics. Two main fields can be established in this sector: on the one hand universities and scientists used programmable machines to solve calculating or logical problems; on the other hand punch card operators and companies used alphanumerical data that was relative to accountancy and the management of mechanical procedures. These processes were limited to punching holes into paper banners and the usage of punched hole cards. The high density of electromechanical equipment used by machines was not very reliable and led to the creation of a new profession. For 35 years maintenance of materials was an essential part of the development of IT systems.

> In 1946 Von Neuman's report defined what should be the electronic calculating machine of the future. A summer school took place during that same year in the University of Pennsylvania. These two facts lead to the development of the first computer program (or software) that was based on punched hole cards and revolutionized the way in which machines processed information.

Over the years, developments in electronics, integrated circuits and punched hole cards have led to an increase in the level of electronics used in machines leading to entirely electronic machines. As their speed and their reliability continually increased, the cost for processing data as well as maintenance continually decreased. However, the main idea which was based on the system of punched hole cards remained the same and there were no profound changes when it came to operating modes.

At the beginning of the 1980s, after experiencing two oil crises the industry rationalized its activities. IT, a support (back office) activity, was not affected by this change. IT professions remained general, i.e. not very specialized, and the majority of people working in the industry generally moved from punched hole cards to magnetic tapes and later on to magnetic disks. Through training and retraining, people working in IT have nearly all come from the era of punched hole cards and their job was reduced to knowing the working of these machines. Developments in the architecture of computers imposing the logic of business clearly exceeded their skills. Due to the belief that this costly measure would be beneficial, the introduction of this type of computer into several different domains and professions led to disillusionment, not to say disappointment, amongst the employees. The systems of the time often restrained employees rather than offering objective advantages. The situation needed to be changed urgently. IT needed to do more than just be a center for simple data processing. At the beginning of the 1980s companies recruited more and more computer scientists that had graduated from university or other prestigious colleges for engineering (in France these are referred to as *grandes écoles d'ingénieurs*). Information systems needed to be adapted to changes and developments in companies and market changes. The director of the IT department is now found on organizational charts. He/she does not only have to ensure the functioning of IT but also take on many other responsibilities. The director of the IT department needs to have a strategic overview of technologies, human resources and how they can be linked to other employees in order to create an ergonomic work environment that is adapted to the user's needs and allows for effective communication. Now part of the board of directors, they are now not only responsible for the running of the company's IT system, but also its development.

This reform is happening on two different levels. Transaction processing systems which enable the person that created a certain type of information to consult it on the system emerged in the 1980s. This system progressively replaced punched hole cards and the employees that carried this task out have been retained as data entry operators. Software developers have replaced those in charge of the team of punch hole card operators and companies increasingly rely on external IT consultants when it comes to organizational functions. This era of IT could be subdivided into six domains: software developers who fulfill two of the functions at the heart of the development team, creating and maintaining the applications throughout their use; database administrators, as database management is a process which is independent of programming; system engineers, whose role grew the more applications shared material, software and data resources; network engineers, in charge of the creation, management and optimization of telecommunications; and, organizational consultants, who translated the needs of the users into specifications and proposed understandable and accessible solutions to problems.

> The profession of engineers working as consultants was created at the end of the 19th century with the emergence of modern industry. For a long time it had been amoral to draw profit from science which represented a great obstacle in the creation of this job profile and is the reason why this type of profession was not developed any earlier. Some of the biggest consultant companies in the field of engineering were founded in the 1920s (in 1926 James O. McKinsey, a professor for accountancy at the Northwestern University, founded McKinsey & Co. and Edwin Booz hired his third employee called Jim Allen and founded Booz Allen & Hamilton). The number of companies working as management consultants increased rapidly during the 1930s. As this time period was an extremely difficult one for the private economic sector the demand for this kind of service increased.

> After WWII the rapid development of technologies and an economic upturn were reflected in ever more complex organizations that increasingly required consultants. In the 1960s the first companies specializing in consulting entered the stock market as a lack of free capital often hindered their development. In 1963 Bruce Henderson founded the Boston Consulting Group which is seen as the first company entirely based on strategy. In 1968 two companies employed more than 1,000 people: Planning Research Corporation (3,000) and Booz Allen & Hamilton (1,500). The profession started to develop.
>
> The economic downturn in the 1970s did not affect the prosperity of consulting companies. The "computerization" of companies, started by the arrival of IBM/360 IT in the previous decade, led to prosperity for consulting companies.
>
> The 1980s were the golden age of consulting companies. The volume more than tripled during this period of good economic growth rate. Consultants, who previously only represented a marginal and quite obscure part of the industry and relatively small companies, from then on were involved in all possible companies or organizations whatever their size or type of activity. Consultant companies from then on started employing very high profile university graduates (e.g. MBA, prestigious universities such as French *grandes écoles* etc.) in order to create an indispensable link with their clients. This link enables software engineers to obtain the information given by the client which is required when creating different sorts of specific applications and meeting the clients needs.

In the 1980s computer scientists who graduated from university and therefore acquired a certain amount of scientific logic were enthusiastic about the potential of new databases especially those associated with matrix-based calculations and linear algebra. Being convinced of these projects this new generation of computer scientists therefore reformed the traditional approach of processing chains and replaced it by elementary data logic. This was the period of functional modelization (SADT, Merise) and the methods that are associated with it. The company's users perceived this systematic approach as a mixture of abstract theory and esoteric neologism which finally lead to IT being isolated within an ivory tower.

In the middle of the 1980s micro-computers were introduced as a competing system within companies. This is a simpler form of IT. Old fashioned managers and employees referred to it as office automation as opposed to "real" IT. Good performances at a low cost, which authorized these new applications (5 to 20 times less costly) that were developed on micro-computers and shared on a network in order to reply directly to the user's needs, have not improved the relationship with the user and traditional IT departments, but led to the creation of new services in the field which is traditionally described as IT. This alternative form of IT has led to the creation of new professions such as engineers specialized in office automation, parc-micro administrators and managers as well as teachers in the field of office automation.

The rise of local networks at the end of the 1980s has started to unify these two IT domains. True computer scientists can use their skills by installing and administrating networks and large data bases shared by several companies which become complex very quickly and easily. Specialists in microcomputing continue to offer an efficient and reactive service in the fields of software development, application management and maintenance of applications such as computer graphics or the processing of multimedia files.

The beginning of the 1990s marked the start of distributed IT and the client/server model. For the first time computing power is shared to a large extent and computers communicate with one another through the company's local network.

> In the beginning IT was centralized. At the beginning of the 1970s the idea of distributed IT in the form of the client/server model emerged for the first time. In reality this is a rather centralized model that is based on a single server that connects to the client. These changes could therefore be referred to as superficial or even "cosmetic" as they only apply when a connection with the client is established.
>
> With the current wave of ever smaller computers, the computers used now are smaller and less powerful than their predecessors, the so-called mainframes. On the other hand smaller computers are therefore forced to co-operate. Micro-computers need to be interlinked with one another in order for them to share their information.
>
> Providers in the field of telecommunications use and design public data networks that interlink these small computers with one another. At this point Transpac and the X25 standard emerged. However, X25 uses the same model used for communication via a telephone. A circuit is connected for the dialog and needs to be disconnected at the end of the conversation.
>
> In the client/server model, where interaction only takes the time required to give an answer to a question asked, no form of dialog is established. The model of a conversation can therefore not be applied at all. Some US universities thus tried to create a different type of network adapted to the model that asks for a specific question. However, this time it should be done without a direct connection. In this type of network there is no preliminary routing process. For every packet the switch has to make a new decision concerning the routing process. This is why switches without a storage facility (i.e. *stateless*) are called routers.
>
> This network is the origin of the Internet as we know it today and its success lies in the fact that the client/server model had first been applied to mini-computers located within companies and then generalized and applied to micro-computers.

The first wave of resource distribution coordinated by a network led to the creation of new professions in the IT sector, especially in the fields of support and providing assistance to clients.

This IT generation also changed in the field of applications where co-operative and even more so collaborative modes allowed for communication and collaboration between different companies (see section 6.5). The communication between different processes produced a considerable amount of information and data structures that evolved much faster than had been the case in the past (information known as "static" in the data schemas used in the 1970s and 1980s became less and less frequent). The life-cycle of applications and the management of different versions that did not concern medium-size companies or organizations and hardly ever large companies or organizations became uncontrollable strains that had to be dealt with.

IT was formally organized around two main "factories" in the company. The "Solution Delivery" (e.g. *Design & Build*) was responsible for the development, integration and maintenance of applications while the "Service Delivery" (e.g. *Run)* was in charge of the use of systems, networks and customer services organized around a call center or *Service Desk* created to capitalize on incidents and offer solutions to these issues.

In a roundabout way we entered the industrial era of distributed IT where functions are specialized and the workforce is largely sub-divided into teams which possess the necessary critical mass to provide for the customer's needs in an efficient and economic way.

At the end of the 1990s or more precisely at the beginning of the 21st century a new era began which is known as web solutions. In the course of the 1990s the Internet developed even further and underwent a normalization process, notably in the fields of messaging systems, intranets[22], company home pages (used for promotion of the company's activities) transforming into the universal form of communication between all possible clients and servers. The notion of thin clients allows for every workstation, no matter how small, to enter a dialog with all possible applications without the previous installation of a specific program. Pervasive IT is now reflecting the expression in its appliances.

---

22 The intranet is a collaborative network that is based on communication techniques such as the Internet Protocol (IP) and used within a company or organization. This tool allows groups of users to access data that is commonly and mutually made available (informational intranet). The second generation of intranet offers access to the company's applications.

Web solutions has furthermore put into question professions in the field of technology and introduced a new socio-professional concept. Organizations and individuals currently co-operate in structured groups such as companies, organizations, teams working on projects, or slightly more informal groupings such as pressure groups. Individuals create, process and share information and knowledge without having to worry about localization of technical strains of a system. Major changes in these less hierarchically structured organizations have altered interactions between individuals as well as interactions between humans and machines. New challenges have emerged in the field of systems as well as in the field of organizations. The social dimension, as well as the geographical spread of players, knowledge and the heterogenity of resources, now need to be taken into account. Furthermore resources need to be validated, verified and certified; this task needs to be managed. Other challenges are the enormous amount of data that needs to be processed and security when accessing data. In the case of individuals the human/machine interface therefore needs to be personalized and adapted to the interaction context and the user's profile (i.e. who can access what kind of data).

The impact on professions in the IT sector is thus very high. This time the impact will be much stronger than was the case with the introduction of micro-computers and multimedia systems. Even the last bastions of the very few mainframe applications that are left have started to become obsolete. Service-oriented architecture (SOA) or more precisely web service programs, are written in many different programming languages and work on different platforms so that they can exchange data via the Internet. Introducing a specific application in a company no longer requires programming skills but rather consists of "consuming" or "using" the services that are available on the Internet for building the desired business process. This new approach is based on assembling components that interact[23] with one another. These components, capable of increasing their activities, show the software developer how they have to be used. The "user manual" is directly integrated into these components. This approach will transform the role and the activities of the IT department. These new services which are available online as opposed to what up till now had been online information will allow for a rapid and comfortable construction of applications without necessarily dealing with functional and technical details in the field of business logic. Using these components not only makes the task of programming easier but also leads to a higher reactivity during the construction process. Maintenance is also affected by the usage of these components because it simplifies the task in itself. Furthermore the usage of this macroscopic service dealing with business processes enables the user to automatically benefit of the developments and improvements made in the individual component.

---

23 Interoperability is the capacity of several systems no matter whether they are identical or very different to communicate effectively and co-operate with one another.

Computer scientists have therefore transformed themselves into users, i.e. consumers of a certain service. They are clients, who choose and later buy the computer's processing capacity, its storage capacity, its bandwidth for telecommunication, access to the components used to write programs and last but not least, and only very recently, also have an impact on business logic. Their job consists of paying attention to and finding the best solutions that exist in a certain domain in order to develop applications that work efficiently for all of the processes within a company or organization. They also have to efficiently interact with external processes (clients, suppliers, stakeholders[24], etc.). The notion of competence in the field of coding has progressively moved on to integrating components. Software developers or in other words "authors" or "clients" now need to observe technological developments, the system architecture and above all analyze the needs of business. These tasks have become part of the requirements software developers need to meet.

To pretend that writing a basic code has entirely disappeared from the profession of software development would be wrong. The components and other web services need to be developed further, tested and maintained in order to ensure their functioning. However, these activities have now been taken over by so-called "component factories", which are highly industrialized and work automatically in order to benefit from a maximum mutualization. It is important to take into account that today no craftsmen, no matter how talented they are, produce their own light bulbs, screws or tools.

The managers of this new IT function are also capable of looking forward to a technological future where the focus will not fundamentally lie on PCs and servers. Managing directors and the company expect directors of the IT department and their staff (e.g. their certified architects) to create new forms of architecture and roadmaps that describe them. These often very radical ideas can be used in the upcoming decade as they are often based on a wide range of technology of which the majority is currently still being created. Furthermore the software developers, now managers of this future technological function are expected to attract users around these new technologies and the behavior they bring about. Their job consists of the promotion of an ambitious dream that to some extent might become real. The technology itself will bring about change. However, managers and administrators of this technology play an important role in the successful outcome of the current development.

Ambient intelligence leads to the convergence of several different disciplines which is therefore referred to as fusion technology. Traditional IT, microsystems, nanotechnologies and molecular electronics make up the next chapter in the history

---

24 A stakeholder is a player from inside or outside a company and is concerned with the company's performance and the fact that it functions well.

of technology where the PC or mainframes are finally replaced by minuscule integrated systems that interact with one another.

Which activities will remain part of the company and on which activities will business set its heart? Professions such as system architects and project managers which are able to analyze the user's needs and combine them with the best possible usage of available technologies will also be working in modern companies. These professions will not be subject to change. However, the increasing interdisciplinarity of IT requires players in the field to broaden their horizons and their field of competence which means that they also have to take other technological and scientific fields into account and gain an overview over this rather large field of science and technology. From a pioneering period[25] onwards a new generation of engineers needs to be trained and recruited, who at the same time need knowledge of IT basics and to be competent in physical domains that deal with fragmented extremely small hardware or capillary platforms. These engineers need to have a broad knowledge in order to be able to integrate this technology in different environments and control it once it is put in place.

RFID applications have been covered to a great extent in this book, not because the basis of these applications is an entirely new approach, but because we believe that this new discipline will be at the origin of the major changes in IT professions that will be experienced by companies. In contrast to traditional IT that tries to unite the interest of two parties (the users and their needs and IT with its solution available on the market) diffuse informatics introduces a third component, a profession that responds to the strains and the peculiarities of the environment in which the respective technology is integrated. This profession is based on the engineer's scientific and technical know-how. This field of IT is far from being new but has been rather specialized and limited to the field of embedded systems (aeronautics, space travel, defense and more recently also in the car industry).

This new generation of IT is, however, the next step into the future. It is used in the latest operational systems that are based on diffuse systems which are embedded into their environment. Whether it is RFID, embedded microsystems, digital mobile device networks or lab-on-chip applications, all economic sectors have already entered the era of ambient intelligence. It is therefore important to establish an overview of professions that will disappear and others that will emerge. This overview could be used as a guideline for university students and help to create curricula that are adapted to the future needs and development of present and future collaborating parties.

---

[25] It is likely that in the long term tools and products that the market offers will allow for a new usage that is more or less free from material strains.

Conclusion

Due to the on-going progress that has been made in the last 50 years, the industry which started off producing electronic components has created information systems and society has benefited from them. The transformation of applications has taken place step by step, but was marked by the technology of computers whose development was relatively stable and at the same time subject to Moore's Law.

With miniaturization, we are currently reaching the boundaries of these technologies. In the next 10 to 15 years, the era of molecular components will begin. This type of device will profoundly change the nature of our computers. This trend will be pushed even further by developments in the field of silicon. Until now, progress was mainly made on centralized units of computers which remained relatively homogenous (e.g. mainframe, PC, server). The analysis of the devices that will be created in the upcoming decade has shown that progress will go hand in hand with a large diversity of capillary systems that co-operate with one another in the form of a swarm of intelligent objects.

Even supercomputers, which are the last bastions of centralized IT based around a monolithic intelligence, are moving towards a system sub-divided into clusters. The computer is transforming into a system. These systems can no longer be reduced to one specialized machine but are spread out in our environment and offer a collective and a co-operative mode. Radical changes in IT will lead to a transformation process in the architecture of information systems which will then mark the boundaries of the future's information technology. Information systems will no longer be centralized in the form of a PC and its user, but will take the shape of a swarm of small communicating devices out of which each single one fulfils its assigned task. This is therefore a complete change. The change will not only affect the way such a system looks, but also the way in which the user perceives it.

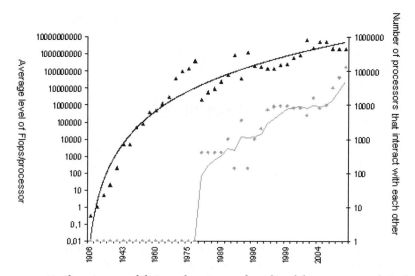

Supercomputers have increased their performance and combined the progress made in the field of processors (silicon technology) which allowed for an increase in the number of processors that co-operate in one computer (architecture systems). The first supercomputers were computers equipped with monoprocessors and were 10 times faster than other computers used at the time. In the 1970s, the majority of supercomputers had adopted a vector processor that carried the decoding of an instruction at once by applying a series of operands. At the end of the 1980s, parallel systems were introduced with the usage of thousands of processors in the same computer. Even today, some parallel supercomputers use RISC microprocessors that are designed for serial computers. Recently, the notion of the supercomputer has changed to the concept of the supersystem which consists of small distributed computers. Riken's MDGrape-3 is the first distributed super system that consists of a cluster of 201 units that can be sub-divided into 40,330 processors. It is the first system to breach the limit of PetaFLOPS (10^{15} FLOPS)

Over the course of one entire century the notion of an automat, a calculator and later a computer, has transformed itself into the universal model of the PC. At the beginning of the 21st century a reverse trend set in. Computers have once again become specialized to carry out their tasks more efficiently. In order to do so their performance is limited to this task only and they can therefore be integrated into the real world. Nanotechnologies allow for further miniaturization and therefore contribute to a successful outcome.

This book would be rather superficial if it only mentioned the development of information systems in the next decade. Why is it important to present so many devices which are currently still emerging or are still being experimented on? We are not summarizing all possible scenarios of the future generation of IT for the sake of it. It is a way of becoming aware of the impact that changes in the field of computers will have on humans who install and manage them.

In order to ensure for a better integration of computers, large appliances will be replaced by specialized digital devices. This specialization and integration in the device's environment requires engineers with new skills and a new breed of computer scientists who are less focused on the creative development of programming code since it will have become universal and written in exchangeable components, but that will focus on the integration of the device into its environment. The trend from specialized to universal is now starting to reverse. At the beginning, the aim was to integrate devices and applications into their environment. With the increasing usage of computers, the IT professions have become more and more abstract and focused on the processing of essentially logical applications. This is the paradigm of plug and play. When connecting a system they configure themselves automatically and are immediately available for tasks that are considered of high importance. Integrating devices directly was no longer considered as a strategic move for companies. When necessary, the concept of sub-contracting workers or companies was applied. IT jobs disappeared just as was the case for internal air-conditioning maintenance teams or car fleet maintenance departments.

RFID and other mobile networks will bring about similar changes. The competencies of higher integration[1] are often delayed or even ignored by the managing directors, which leads to loss in the potential advantage a company would have over its competitors.

This is why companies need to be aware of this tendency and address it when training and hiring people. Students and future computer scientists who are entering the job market also pay a lot of attention to new requirements. During the next decade the professions that were entirely forgotten about in the developments of the last 20 years will re-emerge.

However, the most difficult task will have to be carried out by managing directors, directors of the IT department, CIOs (*Chief Information Officers*) and CTOs (*Chief Technical Officers*). The company's managing directors need to anticipate future developments and plan technological infrastructure in order to have a competitive advantage. Managing IT systems and ensuring that they are working correctly has for a long time been the priority for directors of the IT department. This task has, however, lost its importance. Using systems and offering support to their users are no longer strategic elements of a company. These services are progressively out-sourced to shared service centers, i.e. specialized companies that provide customer services. The maintenance of existing applications is developing

---

1 Competences of integration do not substitute integration services that ensure that projects are implemented. However, they are indispensable when it comes to an internal definition of needs and pilot projects. While the production process needs to be delegated, this should never be done in the phase of development.

256   Nanocomputers and Swarm Intelligence

in the same direction. Applicative and corrective maintenance are out-sourced as they no longer represent a strategic advantage.

No matter how we look at it, until now directors of IT departments had relatively little influence on major strategic decisions that were made within a company. These decisions (e.g. ERP, CRM, e-Business) were mainly taken by the board of directors and carried out by the directors of a specific department that would then co-operate with the IT department.

With the emergence of diffuse IT, the role of technological managers has become far more strategic[2]. Due to the high number and the diversity of omnipresent systems, ambient intelligence will produce a quantity of data that has no equivalent throughout history. All directing boards agree on the fact that the secret of success lies within these enormous amounts of data.

The pharmaceutical industry, for example, produces gigantic amounts of data[3] when a new medication is undergoing preclinical, and later, clinical tests. Research and development of a molecule was until now based on heuristic principles that led to the commercialization of only 1 out of 5,000 tested molecules. It took 12 years for a molecule to be tested in the laboratory and then to finally be launched on the market. 20% of those companies' turnover is invested in R&D facilities.

A new, and by far more rational, approach in R&D is based on being aware of specific chemical reactions of the target organism which will allow the direct creation of the desired molecule. This new technique will have a strong impact on the technology of the next decade. The approach is based on a tool that extracts the required information from a database produced by systems that allow for a very high speed, such as genomic applications[4].

---

2 This cannot be compared to the web economy which was founded at the end of the 1990s as at that point the bases of the profession were not yet fully developed.

3 New forms of embedded systems (e.g. lab-on-a-chip, microsensors that can be used to monitor from a distance) working inside the body, or online (medical imaging) generate the highest amounts of data in the industry.

4 Content-based research is used in very large and unstructured databases. These databases are currently growing so quickly that an increase in the performance of traditional IT still does not ensure that the time needed to obtain results will remain reasonable. In reconfigurable parallel architecture these databases are stored on a high number of hard disks that allow for parallel access. The data is filtered at the exit of the hard disk in order to use only a small amount of data in the calculating process which reduces the time needed to obtain a result. Content based research can start with a simple filter before carrying out a complex and time consuming algorithm. These filters are implemented in the material under the form of reconfigurable components which allow them to be adapted to the format of the data and work in parallel in order speed up the algorithms.

Conclusion 257

Microsystems which are based on nanodevices that obtain data in exactly the same place where the medication should act, i.e. inside the organism, will lead to an exponential increase in the quantity of data.

All data is considered as useful and worthy of being exploited systematically. Never before have IT managers had to process similar amounts of information. A piece of information can be used to create a specific structure and lead to the development of a large variety of devices.

Furthermore, computer scientists are working with programs of a limited lifespan (i.e. 12 years). The emergence of nanocomputing and ambient intelligence are predicted for the near future. Strategic decisions therefore need to be based on upcoming technologies.

> 8 to 12 years are required to develop medication and finally sell it on the market. Nearly an entire decade is needed to conclude all phases of research and development. R&D programs cost several hundred million euros. The process is mainly empirical and based on trial and error. The first step is called physiopathology and defines the target which needs to be addressed in order to cure a disease. Once the target is identified, scientists move on to the conception, the synthesis and finally the test phase of the molecule that will interact with the target. Metaphorically speaking, this can be compared to the key/lock principle. The lock is the target. The medication therefore is the key. The process is subject to the following criteria: the target needs to be addressed effectively without creating any negative side effects.
>
> Choosing and validating the molecular target, for diagnosis and therapeutic reasons, could become much easier when using cheminformatics[5] which allow for a virtual and rapid selection of structures, predictive molecular toxicology, etc. Furthermore, the integration of bioinformatics and cheminformatics in the process of animal testing would be beneficial.
>
> The deciphering of the human genome and those of other organisms provides scientists and medical researchers with a considerable quantity of data. High speed IT systems would be required to exploit this data and use it for a more rational approach towards medical innovations. Information systems that dispose of enormous databases and analyzing tools would allow the most promising molecules to be synthesized and tested in a realistic set-up.

Nearly all other sectors are experiencing the same revolution in the field of IT. Banks and retail, for example, use all different kinds of new systems (e.g. chipcards, radiotags, online payment, mobile payment via cell phone, eye tracking analysis,

---

5 Also known as chemoinformatics or chemical informatics.

etc.) that create huge amounts of data about their clients which might give the company a competitive advantage. Filtering and processing this information is, therefore, of great importance. This is a new vision of IT, as now also data, which is not useful, is produced.

Directors of IT departments therefore need to be able predict future developments. They need to leave behind well known territories which represent monolithic IT and start exploring the areas of capillary technologies. They need to invent solutions for all different sorts of communicating devices. Information will then be sent directly form the place where it is produced (radio tags when moving through doors, digital cardiac pacemakers, geo-localized cell phones, RFID chips, etc.) to its final destination (screen embedded in a vehicle, intelligent watch, etc.). These applications will be increasingly distributed and introduce more and more hybrid functions.

New IT managing directors need to anticipate these developments and use them in their strategies. In other words, they need to create a roadmap that organizes the lifespan of applications in order to suit technological developments as well as the company's human and financial resources. Projects need to be created with the aim of adding value to the company.

Managing directors dealing with the third generation of information systems do not only have to imagine efficient solutions for an era of IT that is currently changing, but also share their enthusiasm with users, heads of department, clients, suppliers and stakeholders[6].

Until now, directors of IT departments were believed to have reached the summit when it comes to power and influence within a company. They are using techniques that have been proven successful in other sectors in the market. Their skills lay in the fields of stabilizing the company, decreasing risks and adapting systems to the required needs by using elementary components in programming code. This was the era of directors who were interested in system operations and tactics. However, we are currently entering a new era. This era will be that of directors that deal with information in a very strategic way. Digital information will be created by all different kinds of devices. The amount of information that will be generated is currently unimaginable. The new breed of IT directors will be characterized by their ability to think abstractly and anticipate and take on calculated risks.

---

[6] There are still too many companies in which the board of directors does not take into account the challenges of technological projects. They also have to enjoy technology and predict future developments in the field in order to be able to admit that the future lies within them. Resources need to become part of their strategies. Directors of IT departments have to help the managing board enjoy technology and predict future developments.

Bibliography

ASSOCIATION FRANÇAISE CONTRE LES MYOPATHIES, *DECRYPTHON project*, AFM, 2002.

AUFFRAY J.P., *Le monde des bactéries, Regard du physicien*, Le Pommier-Fayard, 2000.

BAQUIAST J.P., JACQUEMIN C., *Un ordinateur quantique en 2020?*, Automates Intelligents, 2004.

BENLAHSEN M., *Nanotechnologies: Etat de l'Art*, Université de Picardie Jules Verne, 2000.

BIRGE R.R., GILLESPIE N.B., IZAGUIRRE E.W., KUSNETZOW A., LAWRENCE A.F., SINGH D., SONG Q.W., SCHMIDT E., STUART J.A., SEETHARAMAN S., WISE K.J., "Biomolecular electronics: protein-based associative processors and volumetric memories", *Journal of Physical Chemistry*, 1999.

BIRGE R. R., PARSONS B., SONG Q.W., TALLENT J.R., *Protein-based three-dimensional memories and associative processors*, M.A. Ratner and J. Jortner. Oxford, UK: Blackwell Science, Ltd, 1997.

BRACK A., *Et la matière devint vivante*, Le Pommier, 2004.

CHAKER M., *Les Nanotechnologies: de la manipulation des atomes aux machines nanostructurées*, Colloque Sciences et Technologies, 2000.

CMP CIENTÍFICA, *Nanotechnology: The Tiny Revolution*, CMP Científica, 2001.

DE ROSNAY J., *La spintronique: l'électronique du futur*, Le carrefour du Futur, 2002.

DESSALLES J.L., *L'ordinateur génétique*, Hermes, 1996.

DIGIPLAN TEC INC., *Profil de la main d'œuvre et de l'industrie de la micro-électronique*, Table métropolitaine de Montréal, 2002.

DREXLER K. E., *Engines of Creation: The Coming Era of Nanotechnology*, Anchor Books, New York, 1986.

DREXLER K. E., *Machine-Phase Technology*, Scientific American, 2001.

DREXLER K. E., *Nanosystems: Molecular Machinery, Manufacturing, and Computation*, Wiley-Interscience, 1992.

ESTEVE D., *Le quantronium, premier pas vers un processeur quantique*, CEA Technologies n°63, 2002.

ESTEVE D., *Vers des processeurs quantiques*, Clés CEA n°52, 2005.

FAESSLER J. D., *Les ressources de la grille*, IB-com, 2002.

FERT A., speech given by Albert Fert on the occasion of receiving the gold medal of the CNRS in 2003, 2003

FONDS NATIONAL SUISSE DE LA RECHERCHE SCIENTIFIQUE, *Le futur des télécommunications? Des réseaux de nœuds*, Horizons, 2004.

FONDS NATIONAL SUISSE DE LA RECHERCHE SCIENTIFIQUE, *Le futur des télécommunications? Des réseaux de nœuds*, Horizons, 2004.

G. COLLINS P., AVOURIS P., *Nanotubes for Electronics*, Scientific American, 2000.

GAUTIER J., JOACHIM C., POIZAT J. P. , VUILLAUME D., RAUD S., *La micro-électronique du futur aux Etats-Unis*, Ambassade de France aux Etats-Unis, Mission pour la Science et la Technologie, 2001.

GUILLOT A., MEYER J.A., *Des robots doués de vie*, Le Pommier, 2004.

HALADJIAN R., *De l'inéluctabilité du Réseau Pervasif*, Ozone, 2003.

HATON J. P., PIERREL J. M., PERENNOU G., CAELEN J.,.GAUVAIN J. L, *Reconnaissance automatique de la parole*, Dunod, 1991.

HEHN M., *Jonctions tunnel magnétiques*, LPM Nancy, 2001.

HICKEY B.J., MARROWS C.H, GREIG D., MORGAN G.J., *Spintronics and Magnetic Nanostructures*, School of Physics and Astronomy, E.C. Stoner Laboratory, University of Leeds, 2002.

HUBIN M., *Technologie des composants*, 2002.

IBM ZURICH RESEARCH LABORATORY, *The "Millipede" Project – A Nanomechanical AFM-based Data Storage System*, IBM Zurich Research Laboratory, 2002.

K. BECK, *Extreme Programming Explained*, Addison Wesley, 2000.

K. BECK, *Single-Molecule Junction*, IBM Zurich Research Laboratory, 2006.

KASPER J. E., FELLER S. A., *The Hologram Book*, Prentice-Hall, 1985.

LEFEBVRE A., *Comment l'encre électronique va [peut-être] bouleverser le marché*, zdnet.fr 2003.

LEMARTELEUR X., *Traçabilité contre vie Privée*, Juriscom.net, 2004.

LEMOINE P., *La Radio-Identification*, CNIL, 2003.

LEN M., *Memory of the future: two directions*, Digit-Life.com, 1997-2004.

LIEBER C. M., *The Incredible Shrinking Circuit*, Scientific American, 2001

LÖRTSCHER E., CISZEK J. W., TOUR J., RIEL H., *Reversible and Controllable Switching of a Molecule Junction*, IBM Zurich Research Laboratory, 2006.

LYMAN P., H. VARIAN R., *How Much Information?* 2003, SIMS University of California, Berkeley, 2003

MINSKY M., *La société de l'esprit*, Interéditions 1988.

MULLEN K., SCHERF U., *Organic Light Emitting Devices: Synthesis, Properties and Applications*, Wiley, 2006.

PAKOSZ G., *Le stylo digital*, Presence PC, 2003.

REED M. A., TOUR J. M., *Computing With Molecules*, Scientific American, 2001.

ROUIS J., *L'encre et le papier deviennent électroniques*, EFPG/CERIG, 1999

ROUKES M. L., *Plenty of Room Indeed*, Scientific American, 2002.

RYSER S., WEBER M., *Génie génétique*, Editiones Roche, Bâle, 1992.

SAUNIER C., *Sur l'évolution du secteur des semi-conducteurs et ses liens avec les micro et nanotechnologies*, Sénat – Rapport de l'OPECST n° 138, 2002.

SHANLEY S., *Introduction to Carbon Allotropes*, School of Chemistry, University of Bristol, 2003.

SMALLEY R. E., DREXLER K. E., *Drexler and Smalley make the case for and against "Molecular Assemblers"*, Chemical & Engineering News, American Chemical Society, 2003

STIX G., *Little Big Science*, Scientific American, 2001.

WATERHOUSE COOPERS P., *Navigating the Future of Software: 2002-2004*, PriceWaterhouseCoopers Technology Center, 2002.

WATERHOUSE COOPERS P., *Technology Forecast: 2000 – From Atoms to Systems: A Perspective on Technology*, PriceWaterhouseCoopers Technology Center, 2000.

WHITESIDES G. M., CHRISTOPHER J. LOVE, *The Art of Building Small*, Scientific American, 2001.

WIND S., APPENZELLER J., MARTEL R., DERYCKE V., AVOURIS P., IBM'S T.J. WATSON RESEARCH CENTER, "Vertical scaling of carbon nanotube field-effect transistors using top gate electrodes", *Journal of Applied Physics Letters*, 2002.

YOLE DEVELOPPEMENT, *Étude générique sur les technologies optoélectroniques*, Ministère de l'Economie, des Finances et de l'Industrie, 2002.

Index

A

ABI (Adaptive Brain Interfaces) 176
ADC (analog/digital converter) 108, 166
AFM (atomic force microscope) 43–46, 72, 118, 139, 154, 155
AFM memory 155
Alan Turing 133
allotropy 63
AMD (Advance Micro Devices) 23, 94, 100, 101, 191
AmI (Ambient Intelligence) 138, 185, 200, 232, 235, 238, 239, 241, 244, 251, 252, 256, 257
Andrew Grove 17
ANN (Artificial Neural Network) 167, 168
ASICs (Application-Specific Integrated Circuits) 115
AT&T 108, 124
atomic manipulation 43
ATP (Adenosine TriPhosphate) 158
Auto-ID program 210
automatic speech recognition 165

B

bacteria 47, 56, 60, 61, 76,
bacteriorhodopsin 76, 80, 180
BCI (brain computer interface) 173, 174, 176, 177
Bell Labs 108
bipolar transistor 14, 15
BPDN-DT molecule 144

BrainLab at the University of Georgia 176
Busicom 22

C

carbon allotropes 63
CASPIAN (Consumer Against Supermarket Privacy Invasion And Numbering) 214
Catalysis 53
CEA Quantronics group 129
chipset 22
CISC (Complex Instruction Set Computer) 101, 102, 105
CMM5 (Capability Maturity Model - Level 5) 242, 243
CMOS (complementary metal oxide semiconductor) 14–16, 25, 26, 96, 98, 99, 146, 187
CNFET (Carbon Nanotube Field Effect Transistors) 71, 72
CNT (carbon nanotube) 50, 55, 63–75, 184
connectivity 70, 81, 114, 117
convolution 108
Cornell University 38
CRM (Customer Relationship Management) 110, 153, 256
CTO (Chief Technical Officer) 225
cyclability 145

D

DAC (digital/analog converter) 108
DARPA (Defense Advanced Research Project Agency) 150, 187, 232
data mining 114
datacube 78, 81
DDR RAM (Double Data Rate RAM) 140
decoherence (quantum decoherence) 124, 127
diamondoid 46, 51
distributed computing 115–119
DNA (deoxyribonucleic acid) 47, 47, 56–62, 93, 97, 120, 130–134
DNA computing 130, 132–134
Donald Eigler 38
DRAM (Dynamic Random Access Memory) 30, 104, 140, 143, 145
DSP (Digital Signal Processor) 107–109, 167
dual-core processor 95
Duke University Medical Center 176

E

ecophagy 52
EEG (electroencephalogram) 174–177
electronic paper 164, 165, 177–180
Empedocles of Agrigentum 50
EMS (Event Management System) 213
enzyme 46, 51, 53, 54, 58, 59, 61, 62, 130, 133
EPIC (Explicitly Parallel Instruction Computing) 101, 104, 105
epitaxy 18, 19, 98, 172
Erhard Schweitzer 38, 39
ERP (Enterprise Resource Planning) 33, 91, 110, 153, 166, 202, 256
Escherichia Coli 61
EUV lithography (extreme ultraviolet lithography) 96
eXtreme Programming 236
eye tracking 164, 172, 173, 257

F

Fab 12 25
Fab 36 23

Fairchild 17, 24
Fermi level 10, 68
FET (Field Effect Transistor) 14, 15, 71, 187
FFT (Fast Fourier Transform) 167
Francis Crick 57
fullerenes 51, 63, 64

G

genetic algorithm 130, 231
genetic engineering 46, 55, 56, 60, 61, 132
genome 55, 56, 61, 257
Gerd Binning 42
gesture recognition 170, 171
GMR (Giant Magnetoresistance) 82–84, 148, 149
Gordon Earle Moore 24, 33
Gregor Johann Mendel 56
grid computing 116–118, 226
Grover's algorithm 125
GTIN (Global Trade Item Number) 213
Gyricon 178, 179

H

halobacteria 76
Hamiltonian's Path 131
Harold W. Kroto 63
HCI (human computer interaction) 164
Heiprich Roher 42
Hewlett Packard 105
HMM (Hidden Markov Models) 167
holographic memory 150–153
hybrid mono-molecular electronic circuits 73

I

IBM 27, 29, 38, 39, 42, 70–72, 93, 94, 100, 102, 111, 112, 118, 126, 144, 145, 149, 150, 154–156, 172, 191, 195, 197, 247
IBM BlueGene/L 4
Imation 150
Infineon 145
Institute of Structural Biology (CEA-CNRS-Joseph Fourier University) 76

Intel 4, 17, 22, 24, 25, 29, 31, 93–95, 100–105, 108, 112, 172, 191, 195, 197
Internet of things 209, 211, 212, 213 , 220, 222
interoperability 119, 250
intrication 88, 121, 127
IP (Internet Protocol) 108, 117, 185, 220–224, 249
IPv4 210, 222–224
IPv6 210, 224
Itanium 20, 104, 105
ITRS (International Roadmap for Semi-conductors) 29, 30

J–K

Jack Kilby 17
James Watson 57
John Bardeen 7, 72
Josephson junction 129
Julius Edgar Lilienfeld 14
K. Eric Drexler 42, 45–54

L

lab-on-a-chip 187, 256
Leonard M. Adleman 131, 134
lithium-ion battery 159
lithium-polymer battery 158–160
LMNT (Limited Molecular NanoTechnology) 51
Lucent Technologies 150

M

MAC (Multiply and Accumulate) 107, 108
mass storage 84, 92, 135, 137, 138, 139, 140, 143, 146, 153, 158, 199
MEG (magnetoencephalogram) 174
microsystems 2, 137, 161–163, 187–190, 193, 218, 251, 252, 257
Millipede project 154
MIT (Massachusetts Institute of Technology) 178, 179, 209, 214
MLC Memory (Multi-level Cell Memory) 150
MNT (molecular nanotechnology) 6, 45, 46, 49–52, 54, 60

mobility 113, 136, 157, 218, 219, 221, 223–225, 226, 232
molecular assemblers 49, 50, 54
molecular electronics 70, 74, 75, 251
molecular manufacturing 42, 46
molecular memory 156, 157
mono-atomic transistor 75
Moore's Law 22, 24, 33–35, 107, 122, 149, 195, 232, 253
Motorola 29, 145, 182, 184, 187
MTJ (magnetic tunnel junction) 145
multimedia application 105, 106, 187
mutant algorithm 231
MWNT (multi-walled nanotube) 64–66

N

nanoelectronics 70, 82, 134
nanofactory 54
nanomachine 36, 39, 45–47, 49, 56, 157
nanometer (nm or 10^{-9} m) 5, 6, 23, 25, 26, 28, 30, 38, 45, 47, 52, 64, 87, 93–97, 100, 108, 144, 150, 190, 191, 195
nanorobots 46, 52
nanotechnology 5, 6, 36–38, 42, 45, 49, 51, 52, 55, 60, 62, 66, 67, 74, 91, 93, 94, 190, 195, 218
National Institute of Standards and Technology (NIST), Berkeley University 130, 187, 232
NEC Fundamental Research Lab, Tsukuba 62, 65
Nobel Prize 38, 42, 51, 66
non-Boolean 91, 120
Northbridge 22

O

offshore development 243
OGSA (Open Grid Services Architecture) 117
OLED (Organic Light-Emitting Diode) 165, 181–184
organic molecular electronics 75
Orwellian universe 8

P

P2P (peer-to-peer) 116, 138, 185, 186
PAP (photo-addressable polymers) 152
PARC (Xerox's Palo Alto Research Center) 3, 178, 200
PDA (personal digital assistant) 94, 109, 114
PDMS (polydimethylsiloxane) 97
pervasive 8, 9, 163, 164, 165, 193, 200, 249
PetaFLOPS 4, 254
Philips 29, 145, 180, 207, 209
photolithography 5, 20, 25, 37, 38, 93, 95–97, 99, 162, 195
PML (Physical Markup Language) 211, 213
PN junction 13, 18
polymerase 61, 132
protein 49, 53, 55, 58–60, 62, 75, 76, 78–81, 93, 97, 118, 137, 156, 157, 159

Q

quantum computing 122
quantum confinement 85
quantum dots 84–88
quantum superposition 122, 123
qubit 84, 121–130

R

radiotag 207, 257
RAMAC (Random Access Method of Accounting and Control) 149
RFID (Radio Frequency IDentification) 204–218, 252, 255, 258
ribosome 49, 54, 57, 59, 60
Rice University, Houston 63, 66
Richard Feynman 122
Richard Smalley 51–54, 66
RIED (Real-time In memory Data capture) 213
Riken MDGrape-3 4, 254
RISC (Reduced Instruction Set Computer) 101, 102, 105, 254
RNA (ribonucleic acid) 49, 59, 93
Robert Birge 75, 78
Robert F. Curl 63

Robert Noyce 17
Rock's Law 24, 29

S

savant concentrator 213
SEM (Scanning Electron Microscopy) 144
Schrödinger's cat 123, 124
SDRAM (Synchronous Dynamic Random Access Memory) 140
self-assembly 45, 46, 48, 97, 133
self-replication 49, 50
semi-conductors 4, 9, 10, 16, 17, 26, 27, 29, 30 65, 68, 69, 81, 84, 94, 120, 130, 139, 140, 142, 145, 146, 157, 158, 162, 189, 193, 194
Shinshu University of Nagano 65
Shor's algorithm 126
signal processing 107, 108
silicon 4, 5, 7, 9–11
silicon wafer 18–22, 25, 94, 95, 99
single-electron memory 143, 144
SLM (spatial light modulator) 79, 151, 152
SOA (service-oriented architecture) 250, 111
SOAP (Simple Object Access Protocol) 117
SOC (system-on-a-chip) 115
SOI (Silicon On Insulator) 98, 99
Sony 110, 145, 207, 209
Southbridge 22
spin valve 82, 84
spin-LED 82
spintronics 82–84, 145, 149
SPM (scanning probe microscope) 38, 39, 47
SRAM (static RAM) 23, 142, 143
Stanford University 65
STMicroelectronics 29
strained silicon 100
supercomputer 3, 109, 110, 112, 115, 117, 118, 119, 253, 254
swarm intelligence 92, 186, 224, 229, 230
SWNT (single-wall nanotube) 64–66

T

Tablet PC 171
Technical University of Graz 177

Texas Instruments 17, 108, 109, 191, 207
TFLOPS 3, 4, 119
thermocompression 22
TMR (tunnel magnetoresistance) 145, 149
TMS (Task Management System) 213
touchscreen panel 171
transconductrance 72
transponder 205, 206
TSMC (Taiwan Semi-conductor Manufacturing Co.) 23, 27, 28

U

ubiquity 1, 165
UDDI (Universal Description, Discovery and Integration) 117
University of Carnegie Mellon 170
University of Karlsruhe 170
University of Leipzig 14
University of Pennsylvania 245
University of Rochester 177
University of Santa Barbara 130
University of Sussex 63
University of Texas 231

V

valence band 10, 11, 69
virtual ants 229
virus 45, 55, 56, 58, 60–62, 191, 230, 232
VLIW (Very Long Instruction Word) 101, 105, 108

W–X

Walter Brattain 7, 72
William Shockley 7, 72
W.M. Keck Institute Center for Molecular Electronics 75
WSDL (Web Services Description Language) 117
X25 248
Xerox 3, 52, 178, 179, 200
X-ray lithography 5, 95